Program Characteristics That Contribute to Cost Growth

A Comparison of Air Force Major Defense Acquisition Programs

Mark A. Lorell, Leslie Adrienne Payne, Karishma R. Mehta

Prepared for the United States Air Force
Approved for public release; distribution unlimited

For more information on this publication, visit www.rand.org/t/RR1761

Library of Congress Cataloging-in-Publication Data is available for this publication
ISBN: 978-0-8330-9710-1

Published by the RAND Corporation, Santa Monica, Calif.

Preface

RAND Project AIR FORCE (PAF) is engaged in a multiyear project—*Weapon System Acquisition and Cost Analysis Umbrella Project*—to conduct analyses of interest to the Deputy Assistant Secretary for Acquisition Integration, Office of the Assistant Secretary of the Air Force for Acquisition (SAF/AQX), to improve weapon system–acquisition outcomes and develop better cost-estimating tools for use by the acquisition community.

As part of that effort in fiscal year 2015, the RAND PAF project team worked on several tasks related to maintaining, updating, and enhancing the RAND PAF Selected Acquisition Reports (SARs) database to facilitate life-cycle cost and schedule analyses in Major Defense Acquisition Programs (MDAPs). This report is a companion report to an earlier report, which identified the main characteristics of six recent U.S. Air Force acquisition programs with extreme cost growth.[1] This report evaluates four recent Air Force MDAPs with low cost growth and compares and contrasts their key characteristics to the six programs evaluated with extreme cost growth from the earlier report.

The purpose is threefold. First, we seek to determine whether or not the key characteristics identified in the programs with extreme cost growth are present in the programs with low cost growth and, if not, why. If those characteristics are not present, we assume that this finding reinforces our view that the key characteristics of the extreme cost-growth programs that were identified are likely the root causes of extreme cost growth. Second, we seek to determine the common characteristics of the low cost-growth programs and whether such characteristics can be incorporated into future Air Force MDAPs. Finally, we revisit the main recommendations from our earlier report regarding approaches to mitigating extreme cost growth and, based on our findings from the low cost-growth programs, determine whether those recommendations are still valid and broadly applicable to future Air Force MDAPs.

This report provides summary case studies of the four MDAPs with low cost growth, how the key characteristics of these programs compare with the six programs with extreme cost growth, and how these findings affect our earlier recommendations on mitigating the causes of extreme cost growth.

RAND Project AIR FORCE

RAND Project AIR FORCE (PAF), a division of the RAND Corporation, is the U.S. Air Force's federally funded research and development center for studies and analyses. PAF provides the Air

---

[1] See Mark A. Lorell, Robert S. Leonard, Abby Doll, *Extreme Cost Growth: Themes from Six U.S. Air Force Major Defense Acquisition Programs*, Santa Monica, Calif.: RAND Corporation, RR-630-AF, 2015.

Force with independent analyses of policy alternatives affecting the development, employment, combat readiness, and support of current and future air, space, and cyber forces. Research is conducted in four programs: Force Modernization and Employment; Manpower, Personnel, and Training; Resource Management; and Strategy and Doctrine.

Additional information about PAF is available on our website:

www.rand.org/paf

This report documents work originally shared with the U.S. Air Force on October 6, 2015. The draft report, issued on December 16, 2015, was reviewed by formal peer reviewers and U.S. Air Force subject-matter experts.

Contents

Figures

Tables

Summary

Introduction and Overview

The RAND Corporation recently analyzed key characteristics of six U.S. Air Force Major Defense Acquisition Programs (MDAPs) that experienced extreme cost growth.[2] The findings of this research were derived from detailed case studies and analysis of Selected Acquisition Report (SAR) data.[3] As a companion to that analysis, this report identifies and characterizes conditions present in four other recent Air Force MDAPs that experienced the lowest cost growth among all recent ACAT I MDAPs. This research was commissioned by the Deputy Assistant Secretary for Acquisition Integration, Office of the Assistant Secretary of the Air Force for Acquisition.[4]

The purpose of this report is to compare and contrast the key attributes of programs with extreme cost growth to those of programs with low cost growth to gain insights into the key cost drivers and root causes of cost growth in MDAPs. This document provides the analysis of the best-performing programs as a "control" set to help determine whether the attributes common to the worst-performing programs are likely the main drivers behind extreme cost growth.[5] More-confident identification of program characteristics that contribute to cost growth can enable such factors to either be avoided or mitigated in early program planning. This is ultimately intended to provide the foundation for analysis that will assist the Air Force in developing improved acquisition policies and procedures. Those will contribute to better program outcomes in the areas of cost, schedule, and performance and help reduce the likelihood of future programs' experiencing extreme cost growth.

For the reader's convenience, we summarize our prior research on the worst-performing recent Air Force MDAPs below. Our selection criterion for these programs was straightforward.

---

[2] We define *extreme cost growth* as a percentage at least one standard deviation above the mean cost growth of all programs in one of five cost-growth categories. See Mark A. Lorell, Robert S. Leonard, Abby Doll, *Extreme Cost Growth: Themes from Six U.S. Air Force Major Defense Acquisition Programs*, Santa Monica, Calif.: RAND Corporation, RR-630-AF, 2015. Our definition is focused on Air Force Acquisition Category (ACAT) I MDAPs, which are large acquisition programs currently defined as those with a dollar value for all increments of the program "estimated by the Defense Acquisition Executive Summary (DAES) to require an eventual total expenditure for research, development, and test and evaluation (RDT&E) of more than $480 million in Fiscal Year (FY) 2014 constant dollars or, for procurement, of more than $2.79 billion in FY 2014 constant dollars" or which are designated as such by the Milestone Decision Authority. See Department of Defense Instruction, 5000.02, *Operation of the Defense Acquisition System*, Washington, D.C., U.S. Department of Defense, January 7, 2015.

[3] SARs are congressionally mandated annual reports on program progress, costs, and schedules for all U.S. Department of Defense (DoD) ACAT I MDAPs. RAND findings were reported in Lorell, Leonard, and Doll, 2015.

[4] The technical director of the Air Force Cost Analysis Agency was the project monitor.

[5] For an overview of statistical approaches to causal analysis, see Guido W. Imbens and Donald B. Rubin, *Causal Inference in Statistics, Social, and Biomedical Sciences*, New York: Cambridge University Press, 2015.

Based on past analyses of MDAP cost growth using hundreds of MDAPs from all three services dating back more than four decades, RAND cost analysts identified the programs whose cost-growth percentage was at least one standard deviation above the mean cost-growth percentage of the total sample. A 2014 RAND companion document analyzed the most-recent cost-growth trends among 30 Air Force MDAPs submitting SARs to Congress over the preceding several years or otherwise of interest to the Air Force.[6] Of these, six programs, or approximately 20 percent, experienced extreme cost growth according to our definition.[7] These programs are as follows, in alphabetical order:

- Advanced Extremely High Frequency (AEHF) satellite system
- C-130 Avionics Modernization Program (AMP)
- Evolved Expendable Launch Vehicle (EELV) program
- RQ-4 Global Hawk high-altitude long-endurance unmanned aerial vehicle (Global Hawk)
- National Polar-Orbiting Operational Environmental Satellite System (NPOESS)
- Space-Based Infrared System High (SBIRS High).

We thoroughly assessed and evaluated these six programs to determine the main causes of extreme cost growth to glean "lessons learned" for future Air Force MDAPs. The RAND research cut-off date for these programs was December 2013. Two of the programs have been canceled (C-130 AMP and NPOESS), but the other four have stabilized and are proceeding forward and performing in a more-favorable fashion. We used the earlier history of these programs to better inform Air Force decisions regarding the structuring of future programs in their early stages, but do not comment on the ultimate outcome and more favorable recent performance of the remaining four.

Our in-depth qualitative analysis of programmatic histories of the six cases with extreme cost growth indicates that two main categories of common characteristics and conditions, comprising five subelements, were prominent in these programs:

- premature approval of Milestone (MS) B[8]
  - insufficient technology maturity and higher integration complexity than anticipated
  - unclear, unstable, or unrealistic requirements

---

[6] The RAND SAR database includes hundreds of MDAPs dating back to the 1960s. Of those reporting at least one SAR, 111 were managed by the Air Force. Only 36 of these have cost estimates at MS B and sufficient SAR data to be statistically useful. Of these, seven are ongoing programs, and the rest are completed or nearly complete. An additional four programs are new-start programs. See Robert S. Leonard and Akilah Wallace, *Air Force Major Defense Acquisition Program Cost Growth Is Driven by Three Space Programs and the F-35A: Fiscal Year 2013 President's Budget Selected Acquisition Reports*, Santa Monica, Calif.: RAND, RR-477-AF, 2014.

[7] Historically, approximately 10 percent of the total programs in the database showed extreme cost growth. The higher percentage with extreme cost growth among the current programs does not necessarily indicate that outcomes of more recent programs are worse than historical programs because of a variety of technical cost-comparison issues. For further discussion, see Leonard and Wallace, 2014.

[8] MS B is the point when a major defense acquisition program formally begins and enters the engineering and manufacturing development phase (formerly full-scale development phase).

- unrealistic cost estimates

- suboptimal acquisition strategies and program structure
  - adoption of acquisition strategies and program structures that lacked adequate processes for managing risk through incrementalism or through provision of appropriate oversight and incentives for the prime contractor
  - use of a combined MS B/C milestone is based on the assumption that little or no RDT&E is required but has often been linked to an underestimation of required development work and often led to excessive concurrency between development and production phases.

These categories and characteristics and their distribution across the six MDAPs with extreme cost growth are summarized in Table S.1.[9]

Based on a review of the roughly 30 programs from the SAR database analytically assessed in a companion 2014 document,[10] we were able to identify only four suitable programs with clearly demonstrable low cost growth for comparison with the six programs with extreme cost growth. These four programs are listed alphabetically:

- C-5 Reliability Enhancement and Re-Engining Program (RERP)
- Joint Direct Attack Munition (JDAM)
- Small Diameter Bomb (SDB) Increment I (SDB I)
- Wideband Global SATCOM [Satellite Communications] (WGS) Block I.

We examined the case histories of the low cost-growth programs and compared them carefully with the case histories of the extreme cost-growth programs. Chapter Two provides abbreviated case histories of the four programs with low cost growth, with a focus on the five main subcategories of issues that were found to be key characteristics of the extreme cost-growth programs in our earlier research. The core questions we sought to answer with our case studies of the four programs with low cost growth were:

- Did these programs possess similar characteristics and experience similar challenges as the extreme cost-growth programs? Why or why not?
- If so, how were they managed successfully?
- What else is different about these programs?

---

[9] As shown in Table S.1, not all programs examined experienced all the characteristics and conditions identified. For a full discussion of each category and characteristic, see Chapter One.

[10] Leonard and Wallace, 2014. For a full explanation of the criteria for choosing these four programs, see Chapter One.

Table S.1. Two Categories of Common Characteristics of Six MDAPs with Extreme Cost Growth

	AEHF	C-130 AMP	EELV	Global Hawk	NPOESS	SBIRS High
Premature MS B						
Immature technology; integration complexity	√	√		√	√	√
Unclear, unstable, or unrealistic requirements	√	√		√	√	√
Unrealistic cost estimates	√	√	√	√	√	√
Acquisition policy and program structure						
Acquisition strategy and program structure not tailored for level of risk	√	√	√	√	√	√
MS B/C	√		√	√	√	
Program Acquisition Unit Cost (PAUC) growth (%)	95	193	273	152	154	279

NOTES: The bottom row shows PAUC growth for each program. There is little or no correlation between the number of characteristics evident in a specific program and the severity of that program's cost growth in percentage terms. Each MDAP is unique in context and circumstances, and this is why it is so important to convey the details of each case history and ultimately to compare these six "worst of the worst" cases with best-performing MDAPs. Approval of an MS B/C can be indicative of excessive overlap of the MS B development phase and MS C production phase or underestimation of necessary RDT&E. Such overlap or underestimation can result in an unstable design at the beginning of low rate initial production, necessitating costly modifications and retrofits later.

Findings and Observations

Our assessment of the four programs with low cost growth raises two basic questions:

- Do these findings tend to confirm that the key characteristics of the six programs with extreme cost growth are indeed the key cost drivers of extreme cost growth?
- To what extent are the acquisition approaches used on the programs with low cost growth applicable to future programs?

The Root Causes of Extreme Cost Growth

Our findings regarding the four programs with low cost growth support the contention that the key characteristics identified on the programs with extreme cost growth are indeed important causes of extreme cost growth and cost growth in general (see Chapter Three). With the possible exception of the WGS All Blocks, the low cost-growth programs exhibit few of the key characteristics of the six extreme cost-growth programs, as shown in Table S.2.[11]

The C-5 RERP shows two of the six characteristics of the extreme cost-growth programs, including underestimation of technical complexity and difficulty and unrealistic initial cost estimates. This is perhaps to be expected because, at 18 percent, it experienced by far the most PAUC growth of the four programs with low cost growth. WGS Block I shows one of the

[11] We focus our WGS analysis on WGS Block I. For a detailed explanation of our methodology, see Chapter Two.

Table S.2. Four Low Cost-Growth MDAPs Have Fewer of the Common Characteristics of the Six MDAPs with Extreme Cost Growth

	C-5 RERP	JDAM	SDB I	WGS Block I Only	WGS All Blocks
Premature MS B					
Immature technology; integration complexity	√				
Unclear, unstable, or unrealistic requirements					
Unrealistic cost estimates	√				√
Acquisition policy and program structure					
Acquisition strategy and program structure not tailored for level of risk					√
MS B/C				√	√
PAUC growth (%)	18	–12	–16	9	48

NOTES: The last row shows PAUC growth for each program. The light gray in the last column notes that, if the two follow-on buys are included (Block II and B2FO), more PAUC growth was experienced (a total of 48 percent). See body of the main report for further explanation.

characteristics of the extreme cost-growth programs, and it had the second-highest cost growth of the four low cost-growth programs. And, if all three WGS blocks are examined, it experienced more cost growth, although not nearly as much as programs with extreme cost growth.[12] And the case studies show that the severity and magnitude of the issues confronted by C-5 RERP regarding technology complexity and cost estimates, as well as WGS Block I with respect to premature MS C, were far less challenging than those experienced on the six programs with extreme cost growth.[13]

In short, our assessment of four MDAPs with low cost growth and comparison with the six MDAPs with extreme cost growth indicate that there appears to be a strong relationship between the program attributes identified in our prior research and extreme cost growth. Conversely, these attributes are found to be largely lacking or of much smaller magnitude in the programs with low cost growth. Therefore, this comparison provides us with much greater confidence that the key characteristics of programs we identified in our earlier research are indeed among the key drivers and root causes of extreme cost growth on recent Air Force MDAPs.

---

[12] Based on the RAND SAR database methodological approach, only WGS Block I was examined to see whether it qualified as a low cost-growth program. See Chapters Two and Three for a full discussion of the WGS cost methodology and approach.

[13] The WGS program led to three separate blocks or versions of the baseline satellite: Block I, Block II, and Block II Follow-On. Only the first block of satellites (Block I) is considered in this analysis. Table S.2 shows the assessment of all the WGS blocks taken together. Chapters Two and Three explain the different outcomes for Block I compared with all WGS blocks and how we interpret them.

Applicability of Findings to Future Air Force MDAPs

Are there other significant attributes common to all the four low cost-growth programs that clearly differentiate them from all the programs with extreme cost growth and thus can be possibly associated with the difference in cost-growth outcomes? Indeed, while each MDAP is unique in some degree, it is apparent that most of the programs with extreme cost growth are much larger in dollar value both in total program and RDT&E costs.

If we examine estimated RDT&E costs for all ten programs at MS B, we see a pronounced difference between the programs with extreme cost growth and those with low cost growth. Figure S.1 compares the MS B RDT&E cost estimates of all the MDAPs with extreme cost growth and low cost growth, as well as the estimates as of the December 2013 SAR (or program completion, if first), in constant 2015 dollars, showing the real cost growth to that date. All of the programs with low cost growth have smaller RDT&E estimates at MS B, with the exception of the C-5 RERP. Three of the four programs with low cost growth have MS B estimates considerably below the lowest MS B RDT&E estimate for an extreme cost-growth program (C-130 AMP). In addition, the best performing of the low cost-growth programs—JDAM and SDB I—have RDT&E estimates at MS B far below those for any of the extreme cost-growth programs. The worst performing of the programs with low cost growth—C-5 RERP—has by far the highest RDT&E estimate at MS B among the low cost-growth programs.

The size of the initial RDT&E cost estimates and the program's subsequent cost-growth performance might be viewed as a rough proxy for technological and/or integration complexity and difficulty. For the most part, our case-study analysis confirms that the six extreme cost-growth programs tended to be much larger, more complex, and challenging development (as well as production) programs than the four programs with low cost growth.

Does this finding—that smaller dollar value of the estimate at MS B and generally less complexity for the MDAPs experiencing low cost growth compared with the extreme cost-growth programs—mean that the lessons learned from the low cost-growth programs are only truly applicable to similar types of lower complexity programs? We think not.

Figure S.1. RDT&E MS B and December 2013 SAR Cost Estimates by Cost-Growth Category

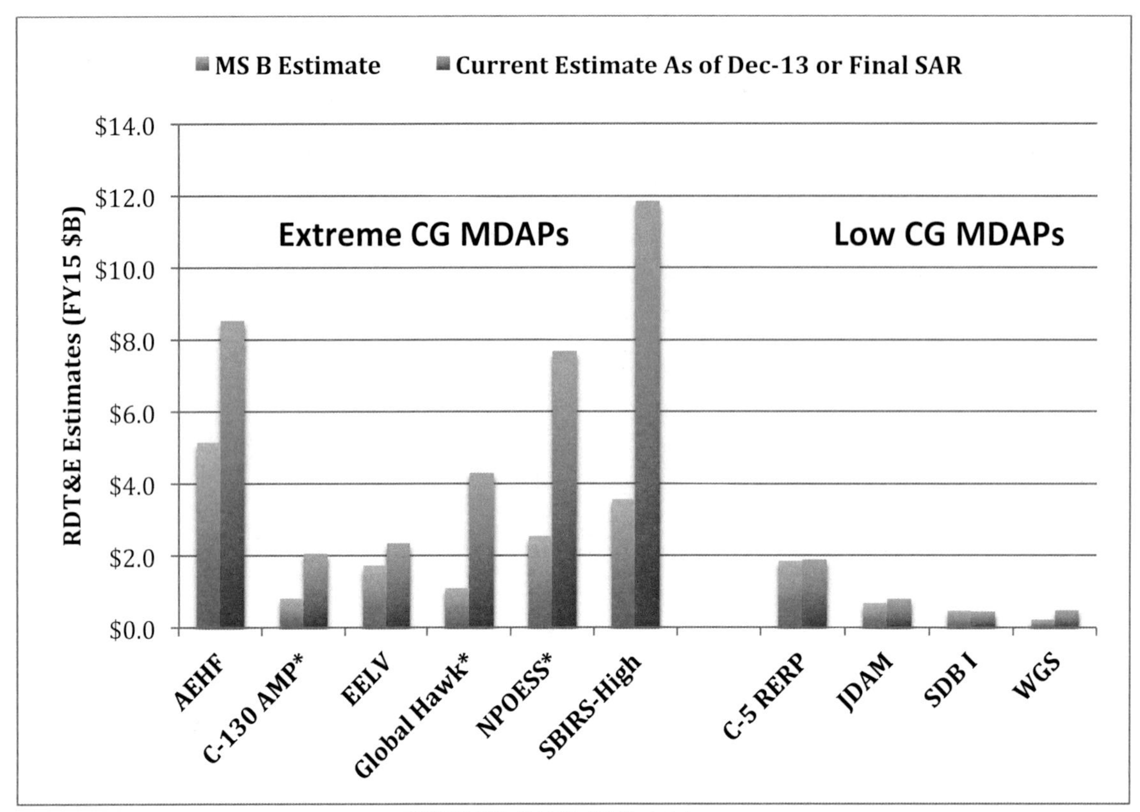

SOURCE: RAND SAR database.
NOTES: CG = cost growth.
* Programs were terminated or truncated in the FY 2013 President's Budget.
Note that three of the four programs with low cost growth have MS B estimates considerably below the lowest MS B RDT&E estimate for an extreme cost-growth program.

Extreme Cost Growth and Possible Mitigation Approaches

We argue that the key characteristics of the programs with extreme cost growth, as summarized in Table S.1, represent the causes of extreme cost growth and apply to all MDAPs no matter how complex and challenging. The near complete absence of the key characteristics of the extreme cost-growth programs from the low cost programs, as shown in Table S.2, confirms our earlier assessment that these characteristics are important root causes of extreme cost growth. What the lower complexity of the low cost-growth programs tells us is that it is easier to apply strategies and approaches to avoid or mitigate these characteristics in less-complex programs than in more-complex programs. However, we hypothesize that a low-complexity program will not necessarily automatically achieve low cost growth without application of corrective or mitigating measures and that similar but more extensive mitigating measures can be also applied to much more complex programs with beneficial effects.

We believe our finding in this research that the low cost programs tend to be less challenging and complex structurally and technologically supports our central recommendation for mitigating extreme cost growth in our earlier research document on extreme growth programs.[14] Our two key recommendations from the earlier report were to

- ensure that programs have realistic cost estimates at MS B
- embrace incremental strategies with comprehensive and proven implementation strategies.

The essence of incremental acquisition strategies is breaking down complex, technologically challenging programs into smaller, less complex, more manageable discrete sequential segments. Our finding that the lower cost-growth MDAPs were all relatively less complex tends to confirm the importance of this strategy. When possible, large, complex programs incorporating cutting-edge technologies and challenging system-integration issues should probably be separated into smaller, less-complex subcomponents, unless urgent requirements or the technological and design configuration of the system make such an approach unfeasible.[15] Without reducing programmatic and technological complexity through the use of smaller-phased sequential segments, large challenging programs may be more likely to experience extreme cost growth. Effective implementation of incremental acquisition strategies can be challenging, particularly in determining the precise content of each increment. Nonetheless, this approach appears to hold out the promise of reducing developmental and integration complexities and risks that may lead to substantial cost growth later in programs.[16] The Air Force needs to continue to experiment with incremental strategies, as well as novel contracting methods and incentives and other approaches to encourage contractors to control cost growth.

---

[14] For a complete discussion of the findings and recommendations of our earlier research, see Lorell, Leonard, and Doll, 2015.

[15] Areas deserving of much more extensive analysis include determining the viability of and best implementation strategies for incremental acquisition. The views expressed here are based on the analysis of the case studies, as well as early RAND research published in Mark A. Lorell, Julia F. Lowell, and Obaid Younossi, *Evolutionary Acquisition: Implementation Challenges for Defense Space Programs*, Santa Monica, Calif.: RAND Corporation, MG-431-AF, 2006.

[16] Each phase of an incremental acquisition strategy aims at producing an operationally useful system. Systems developed in later phases will obviously be more capable than earlier phases or have a different mission focus. The systems developed in earlier phases may be upgraded, retained as lessor capability yet still useful systems, or retired. An example of how this situation is handled can be found in the development and deployment decisions for the RQ-4 Global Hawk program from Block 0 through Block 40. The Air Force eventually chose to retire the earliest systems (Block 0, Block 10, and possibly Block 20), while retaining the more-capable Block 30 and Block 40 systems.

Acknowledgments

We would like to thank Obaid Younossi, director, Resource Management Program, RAND Project AIR FORCE; William Shelton, project leader, Weapon System Acquisition and Cost Analysis Umbrella Project, RAND Project AIR FORCE; and Akilah Wallace, RAND Project AIR FORCE SAR Database project leader, for their support and contributions to this research effort.

Abbreviations

4Q	fourth quarter
ACAT	Air Force Acquisition Category
AEHF	Advanced Extremely High Frequency
AMP	Avionics Modernization Program
APB	acquisition program baseline
APUC	average procurement unit cost
AUPP	average unit production price
B2FO	Block II Follow-On
BY	base year
CAD	competitive advanced design
CAIV	Cost as an Independent Variable
CMI	civil-military integration
COTS	commercial off-the-shelf
CSAF	Chief of Staff of the Air Force
DAES	Defense Acquisition Executive Summary
DAPP	Defense Acquisition Pilot Program
DCSC	Defense Satellite Communications System
DoD	U.S. Department of Defense
DoDI	Department of Defense Instruction
EELV	evolved expendable launch vehicle
EMD	engineering and manufacturing development
FAR	Federal Acquisition Regulation
FFP	firm fixed price
FPEPA	fixed price with economic price adjustment
FRP	full-rate production
FY	fiscal year

GAO	U.S. Government Accountability Office
GE	General Electric
GOTS	government off-the-shelf
GPS	Global Positioning System
IDA	Institute for Defense Analyses
IMU	inertial measurement unit
INS	inertial navigation system
IPT	integrated product team
ISR	intelligence, surveillance, and reconnaissance
JDAM	Joint Direct Attack Munition
JROC	Joint Requirements Oversight Council
KPP	key performance parameters
LRIP	low-rate initial production
MDAP	Major Defense Acquisition Program
MILSPEC	military specification
MS	milestone
NASA	National Aeronautics and Space Administration
NPOESS	National Polar-Orbiting Operational Environmental Satellite System
OSD	Office of the Secretary of Defense
PAUC	Program Acquisition Unit Cost
PBA	price-based acquisition
PPCC	Production Price Commitment Curves
RAA	Required Assets Available
RDT&E	research, development, test, and evaluation
RERP	Reliability Enhancement and Re-Engining Program
RFP	request for proposals
RPV	remotely piloted vehicle
SAR	Selected Acquisition Report
SBIRS High	Space-Based Infrared System High

SDB	small diameter bomb
SDD	system development and demonstration
SOO	statement of objectives
SOW	statement of work
SPO	system program office
SV	space vehicle
TINA	Truth in Negotiations Act
TRA	technology readiness assessment
TSAT	Transformational Satellite System
TSPR	Total System Program Responsibility
USD(AT&L)	Under Secretary of Defense for Acquisition, Technology, and Logistics
WGS	Wideband Global SATCOM

1. Introduction

Overview and Research Approach

The RAND Corporation recently analyzed key characteristics of six U.S. Air Force Major Defense Acquisition Programs (MDAPs) that experienced extreme cost growth.[1] The findings of this research were derived from detailed case studies and analysis of Selected Acquisition Report (SAR) data.[2] As a companion to that analysis, this report identifies and characterizes conditions present in four other recent Air Force MDAPs that experienced the lowest cost growth among all recent Air Force ACAT I MDAPs. Our methodology is to measure cost growth from the formal Milestone (MS) B estimate—the formal beginning of full-scale weapon system development—to the end of the acquisition procurement phase for RDT&E, as well as production (procurement), normalizing for constant dollars and quantity changes.[3]

This report compares the key attributes of the extreme cost-growth programs we found earlier with characteristics of four recent low cost-growth programs to gain insights into the key cost drivers and most-important causes of cost growth in MDAPs. The objective has been to assist the Air Force in identifying and mitigating those characteristics in the earliest MDAP stages that are most likely to lead to substantial future cost growth and to promote those characteristics that appear most closely linked to low cost growth. This report was commissioned by the Deputy Assistant Secretary for Acquisition Integration, Office of the Assistant Secretary of the Air Force for Acquisition.[4]

While RAND's earlier study identified common characteristics and conditions of poorly performing programs, this was not sufficient for clearly identifying the most likely causes of extreme cost growth and their relative importance. To achieve that objective, it was necessary to

---

[1] We define this as a percentage at least one standard deviation above the mean cost growth of all programs in one of five cost-growth categories. Our definition is focused on Acquisition Category I (ACAT 1) MDAPs, which are large acquisition programs currently defined as those with a dollar value for all increments of the program "estimated by the Defense Acquisition Executive Summary (DAES) to require an eventual total expenditure for research, development, and test and evaluation (RDT&E) of more than $480 million in Fiscal Year (FY) 2014 constant dollars or, for procurement, of more than $2.79 billion in FY 2014 constant dollars" or that are designated as such by the Milestone Decision Authority. See Department of Defense Instruction (DoDI), 5000.02, *Operation of the Defense Acquisition System*, Washington, D.C., U.S. Department of Defense, January 7, 2015.

[2] SARs are congressionally mandated annual reports on program progress, costs, and schedules for all U.S. Department of Defense (DoD) ACAT I MDAPs. RAND findings were reported in Mark A. Lorell, Robert S. Leonard, and Abby Doll, *Extreme Cost Growth: Themes from Six U.S. Air Force Major Defense Acquisition Programs*, Santa Monica, Calif.: RAND Corporation, RR-630-AF, 2015.

[3] For more on RAND cost-estimating and cost-growth methodologies, see Robert S. Leonard and Akilah Wallace, *Air Force Major Defense Acquisition Program Cost Growth Is Driven by Three Space Programs and the F-35A: Fiscal Year 2013 President's Budget Selected Acquisition Reports*, Santa Monica, Calif.: RAND, RR-477-AF, 2014.

[4] The technical director of the Air Force Cost Analysis Agency was the project monitor.

compare the worst-performing programs with an approximately equal and similar set of "control" programs that have performed well. The fact that all six poorly performing programs share many common attributes is meaningless if it is found that the same attributes are shared by relatively well-performing programs. Thus, the characterization of the attributes of the six programs in our earlier report was only the first step of our longer-term analytical objective. The second step, the focus of this report, was to identify and assess the key characteristics of the best-performing MDAPs and compare them with the worst-performing MDAPs.[5]

This report provides the analysis of the control set of best-performing programs to help determine whether the attributes common to the worst-performing programs are the true drivers of extreme cost growth. More confident identification of the program characteristics that contribute to cost growth can enable acquisition strategies and program structures that mitigate or avoid such factors. This is ultimately intended to provide the foundation for analysis that will assist the Air Force in developing improved acquisition policies and procedures that will contribute to better program outcomes in the areas of cost, schedule, and performance and help reduce the likelihood of future programs experiencing extreme cost growth.

It is critical to keep in mind that we make all comparisons among programs based on cost growth defined as the growth in Program Acquisition Unit Cost (PAUC) in constant dollars from the MS B baseline estimate. Our calculation of PAUC growth is always normalized for quantities. In other words, no matter what the final quantity procured, whether it be larger or smaller than what was planned at MS B, our calculation of PAUC growth is based on projecting total cost under the assumption that the original quantity planned at MS B was procured. If this is not done, then the PAUC would reflect changes in quantity rather than actual unit cost growth and therefore distort our findings.

Programs may have large changes in total procurement quantities, such as joint direct attack munition (JDAM) or Wideband Global satellite communications (SATCOM) (WGS), which can raise overall budgetary costs dramatically—even leading to Nunn-McCurdy cost-growth breaches—or artificially reduce budgetary costs, although unit cost growth has risen dramatically. That is why it is so important to normalize to the MS B baseline when calculating PAUC.[6]

We used three major sources of data and information to assemble our case histories and qualitative analyses: government documents, published sources, and field research. Almost all the cost data used in our cost-growth analyses were derived from the SARs, which also include annual program summary histories and other details about the programs that are useful for developing case histories.

---

[5] For an overview of statistical approaches to causal analysis, see Guido W. Imbens and Donald B. Rubin, *Causal Inference in Statistics, Social, and Biomedical Sciences*, New York: Cambridge University Press, 2015.

[6] Normalizing RDT&E expenditures to the MS B, baseline is extremely difficult, because it is usually not directly linked to changes in procurement quantity. Therefore, we do not normalize RDT&E for quantity changes.

To provide a more robust foundation for our case studies, we reviewed DAESs for some of these programs. Throughout the life of the program, these reports were typically submitted four times per year.[7] DAESs often provide more and different types of information than the SARs about program issues and events because, unlike SARs, they are generated only for internal use by DoD. They also include comments and insights from multiple sources within the Air Force, as well as within other components of DoD.

We conducted wide-ranging surveys of open-source information on these programs, including such government studies as those from the U.S. Government Accountability Office (GAO),[8] Congressional Research Service, and Congressional Budget Office, along with expert testimony before Congress, studies and dissertations from a variety of DoD and non-DoD sources, and the extensive material available in the trade press, such as *Inside Defense* and *Aviation Week and Space Technology*. This material from outside the official program documentation proved to be crucial for establishing the broader context about many key issues that are relevant to our inquiry.

Finally, in some instances, we were able to conduct interviews with current or former program officials or other subject-matter experts. We also reviewed past documented interviews with senior program officials conducted by one of the authors in support of earlier RAND projects.[9]

For the reader's convenience, the next subsection summarizes the characteristics of the six programs with extreme cost growth that were reported in our earlier document. The current report, however, will otherwise focus on the four MDAPs with low (or no) cost growth.

Six Programs with Extreme Cost Growth[10]

To achieve our research objectives, we built on the earlier RAND work in the area of program cost growth documented in two reports: *Sources of Weapon System Cost Growth: Analysis of 35*

---

[7] DAES usually begin being issued following an MS A or equivalent decision but typically include sparse data and information until the MS B decision time period.

[8] Formerly the U.S. General Accounting Office.

[9] We did not conduct a general literature review on weapon-system cost growth as part of this research because it was outside of the scope of our work effort. However, it is important to acknowledge that a large literature exists in this area, emanating both from the government as well as private sources, including such federally funded research and development centers as RAND. We adopted an approach of focusing on case studies selected through criteria established by our sponsor. We reviewed the general literature when significant information on one of our case studies could be found. One example is a recent RAND report that included an extensive assessment of WGS: Irv Blickstein, Michael Boito, Jeffrey A. Drezner, James Dryden, Kenneth Horn, James G. Kallimani, Martin C. Libicki, Megan McKernan, Roger C. Molander, Charles Nemfakos, Chad J. R. Ohlandt, Caroline Reilly, Rena Rudavsky, Jerry M. Sollinger, Katharine Watkins Webb, and Carolyn Wong, *Root Cause Analyses of Nunn-McCurdy Breaches*, Vol. 1: Zumwalt-*Class Destroyer, Joint Strike Fighter, Longbow Apache, and Wideband Global Satellite*, Santa Monica, Calif.: RAND Corporation, MG-1171/1-OSD, 2011.

[10] This subsection draws directly from Lorell, Leonard, and Doll, 2015.

Major Defense Acquisition Programs (2008) and *Historical Cost Growth of Completed Weapon System Programs* (2006).[11] Based on these prior analyses, the project team concluded that meaningful insights into the root causes of MDAP cost growth required developing in-depth qualitative case studies to supplement the quantitative analysis derived from the SAR data. To achieve the depth of detail needed to attain insight into specific programs, the study team limited the case studies to the worst-performing MDAPs (in terms of cost growth from MS B) over the past several years, which were those of most direct interest and relevance to the Air Force.

Our selection criterion for the worst-performing recent Air Force MDAPs was straightforward. Based on past analyses of MDAP cost growth using hundreds of programs from all three services dating back over four decades, RAND cost analysts defined the programs with the worst cost growth as those with a percentage cost growth at least one standard deviation above the mean cost-growth percentage of the total sample. These programs are described as exhibiting "extreme cost growth." A 2014 RAND companion report analyzed the most-recent cost-growth trends among 30 Air Force MDAPs submitting SARs to Congress over the preceding several years or otherwise of interest to the Air Force.[12] Of these, six programs, or approximately 20 percent, experienced extreme cost growth according to our definition.[13] These programs are as follows, in alphabetical order:

- Advanced Extremely High Frequency (AEHF) satellite system
- C-130 Avionics Modernization Program (AMP)
- Evolved Expendable Launch Vehicle (EELV) program
- RQ-4 Global Hawk high-altitude long-endurance unmanned aerial vehicle (Global Hawk)
- National Polar-Orbiting Operational Environmental Satellite System (NPOESS)
- Space-Based Infrared System High (SBIRS High).

We thoroughly assessed and evaluated these six programs to determine the main causes of extreme cost growth to glean lessons learned for future Air Force MDAPs. The RAND research cut-off date for these programs was December 2013. Two of the programs have been canceled (C-130 AMP and NPOESS), but the other four have stabilized and are proceeding forward and performing in a more-favorable fashion. We used the earlier history of these programs to better

---

[11] Joseph G. Bolten, Robert S. Leonard, Mark V. Arena, Obaid Younossi, and Jerry M. Sollinger, *Sources of Weapon System Cost Growth: Analysis of 35 Major Defense Acquisition Programs*, Santa Monica, Calif.: RAND Corporation, MG-670-AF, 2008; and Mark V. Arena, Robert S. Leonard, Sheila E. Murray, and Obaid Younossi, *Historical Cost Growth of Completed Weapon System Programs*, Santa Monica, Calif.: RAND Corporation, TR-343-AF, 2006.

[12] The RAND SAR database includes hundreds of MDAPs dating back to the 1960s. Of those reporting at least one SAR, 111 were managed by the Air Force. Only 36 of these have cost estimates at MS B and sufficient SAR data to be statistically useful. Of these, seven are ongoing programs, and the rest are completed or nearly complete. An additional four additional programs are new-start programs. See Leonard and Wallace, 2014.

[13] Historically, approximately 10 percent of the total programs in the database showed extreme cost growth. The higher percentage with extreme cost growth among the current programs does not necessarily indicate that outcomes of more recent programs are worse than historical programs, because of a variety of technical cost comparison issues. For further discussion, see Leonard and Wallace, 2014.

inform Air Force decisions regarding the structuring of future programs in their early stages, but do not comment on the ultimate outcome and more favorable recent performance of the remaining four.

As shown in Table 1.1, all six programs experienced extreme cost growth in at least two of the standard five metrics used in RAND's cost-growth analysis. These metrics are grouped under the two broad categories of budgetary and unit cost growth. There are three types of budgetary cost growth: development, procurement, and program.[14] There are also two types of unit cost growth: unit procurement and unit program.[15] The broad budget category is not adjusted for quantity changes in order to show the budgetary effects of program cost growth in constant dollars. However, since it is not adjusted and normalized for quantity change, it is not a good indication of true system cost growth, because planners will often reduce planned procurement quantities of items with high unit cost growth to reduce overall program budgetary cost growth.

Table 1.1. Six Air Force MDAPs with Extreme Cost Growth

Program	MS B or B/C	Budgetary Cost Growth (%)			FY 2015 M$ Growth	Unit Cost Growth (%)	
		Development	Procurement	Program		Procurement	Program
AEHF	November 2001	66	***280***	115	7,600	***190***	94
C-130 AMP*	July 2001	148	24	47	2,000	***194***	***193***
EELV	October 1998	35	***256***	***235***	42,900	***306***	***280***
Global Hawk*	March 2001	***291***	123	157	8,800	84	***152***
NPOESS*	August 2002	106	101	68	4,800	***335***	***154***
SBIRS High	November 1996	***232***	***515***	***300***	14,600	***361***	***266***

NOTES: Percentages shown in bold italics represent extreme cost growth, defined as cost growth more than one standard deviation above the mean for that measure.
* Programs were terminated or truncated in the FY 2013 President's Budget. Cost growth is shown from beginning of the relevant phase through FY 2013: from MS B or B/C for RDT&E, and from MS C or the exercise of the first production contract procurement. This table has been updated from the similar table in our earlier companion report (see Lorell, Leonard, and Doll, 2015).

The last two columns on the right measure unit cost growth from the original MS B estimate and are adjusted for final program quantities, thus providing the best assessment of the original MS B estimate's accuracy.[16] Therefore, we focus on procurement and program unit cost growth

---

[14] Program costs are defined as the sum of development, procurement, military construction, and acquisition-related operations and maintenance costs associated with each program's acquisition.

[15] Unit program cost is defined as total program cost (development, procurement, military construction, and acquisition-related operations, and maintenance) divided by total adjusted program units (quantity normalized to original MS B estimate).

[16] DoDI 5000.2, *Operation of the Defense Acquisition System*, Washington, D.C.: U.S. Department of Defense, May 12, 2003, designates MS B as the formal beginning of an MDAP.

as the best indications of extreme cost growth. Of the 12 unit cost growth measurements for our six systems, ten, by our definition, are extreme (as shown in bold italicized font in Table 1.1), indicating gross underestimates of these programs at MS B.

All six of these programs thus experienced extreme cost growth that, according to our analysis of the SARs, cannot be explained by substantial quantity increases or unforeseeable circumstances taking place beyond MS B which were outside of the program's control. How then can the extreme cost growth of these six MDAPs be explained? What are the true sources and most important drivers of cost growth in these programs? To answer these questions, it was necessary to construct a much more detailed case history of each of them than was possible based solely on the narrative and quantitative data in SARs. We then compared the outcome to the characteristics of the best-performing programs, which served as controls to assist us in identifying the root causes of extreme cost growth. This report presents the best-performing case summaries and the results of that comparison.

Our in-depth qualitative analysis of programmatic histories of the six cases with extreme cost growth indicated that these programs shared two main categories of common characteristics and conditions, comprising five sets of sub-characteristics:

- premature approval of MS B
  - insufficient technology maturity and higher integration complexity than anticipated
  - unclear, unstable, or unrealistic requirements
  - unrealistic cost estimates
- suboptimal acquisition strategies and program structure
  - adoption of acquisition strategies and program structures that lacked adequate processes for managing risk through incrementalism and through provision of appropriate oversight and incentives for the prime contractor
  - use of a combined MS B/C milestone is based on the assumption that little or no RDT&E is required but has often been linked to an underestimation of required development work and often led to excessive concurrency between development and production phases.

These two categories of five sets of characteristics and conditions are summarized in Table 1.2 for all six programs.[17]

Categories and characteristics very similar to those discussed above are commonly found in the acquisition-analysis literature discussing the causes of cost growth and are largely self-explanatory. Our list is not that different from many other similar lists.[18] However, we

---

[17] As shown in Table 1.2, not all programs examined experienced all the characteristics and conditions identified.

[18] For example, the Institute for Defense Analysis (IDA) published comprehensive study findings on the causes of cause growth in 2009 that identified many of the same factors, although the RAND study was conducted without any reference to or knowledge of this study (see Gene Porter, Brian Gladstone, C. Vance Gordon, Nicholas Karvonides, R. Royce Kneece Jr., Jay Mandelbaum, and William D. O'Neil, *The Major Causes of Cost Growth in Defense Acquisition*, Vol. I, *Executive Summary*, Alexandria, Va.: Institute for Defense Analyses, IDA Paper P-

attempted—through the comparison of programs with extreme cost growth to those with little cost growth and through the use of in-depth case studies based on comprehensive interviews with program officials—to provide a more nuanced and prioritized qualitative assessment of these causal factors and to help provide more specific and actionable remedies, compared with most other studies.

Thus, we did two types of evaluation not commonly done in other similar studies. First, we compared and contrasted extreme cost-growth programs to low cost-growth problems. The purposes were to help validate the central importance of the two categories and five characteristics as key causes of extreme cost growth. If we consistently found key characteristics in well-performing programs that were the opposite of these five characteristics, we assumed that this strengthened the evidence supporting these as key characteristics of programs with extreme cost growth. Our comparison did indeed show this, thus confirming our assessment. Second, unlike most discussions of the causes of cost growth, we attempted to prioritize the relative importance of various key causal factors we identified. We did this in a broad, subjective manner by again comparing the poorly performing program characteristics to the well-performing program characteristics.

Finally, although our basic categories and characteristics are not unusual and are essentially self-explanatory, we felt it would be helpful to some readers to providc some additional general discussion of each characteristic within the two categories.[19]

Our first category focused on the general overall readiness of programs to pass through MS B and move on into a successful full-scale development phase free of substantial cost growth. MS B is the official launch point for MDAPs and is the point from which developmental cost growth is measured. In this general category, we found three key characteristics of programs that were clearly not ready for full-scale development and which resulted in extreme cost growth after MS B. In addition, we found the opposite of these characteristics in the programs that performed with little cost growth after MS B.

4531, December 2009a). The IDA report reported two broad categories of problem areas similar to our two broad categories: "Weakness in initial program definition and costing," and "Weakness in management visibility, direction, and oversight." Within the first category, IDA included factors of requirements definition, immature technologies, system engineering, and schedule compression and concurrency. Within the second broad category, IDA discussed problems of lax implementation of policies, which focused on poorly understanding the system requirements and ensuring technology maturity. Inappropriate acquisition approaches are also called out. More recently, the Defense Acquisition University examined five different studies of cost growth by different organizations and found many common factors, several of which are very similar to ours, such as "inadequate analysis of risk" and "estimating errors" (see Allen Friar, *Cost Growth and the Limits of Competition*, Huntsville, Ala.: Defense Acquisition University, September 2012). Indeed, these same themes run throughout the entire modern history of acquisition policy reform, as shown by such studies as J. Ronald Fox, *Defense Acquisition Reform: 1960–2009: An Elusive Goal*, Washington, D.C.: Center of Military History, U.S. Army, 2011.

[19] It is important to note that, although our categories and characteristics are similar to many other analyses of MDAP cost growth, they are still founded to a great degree on qualitative assessment and judgments. Other valid interpretations of the evidence are of course possible and may even have advantages compared with our analysis.

Technology maturity and integration complexity. Our assessment based on comparisons of the worst performing programs to the well performing programs suggests this is the single most important characteristic determining whether or not a program will experience extreme cost growth. Technology maturity refers to the readiness of a specific broadly defined technology or technology area for application within a specific context of a weapon system design and development. The readiness of the technology for incorporation into an MDAP is directly linked to the level of risk involved in developing the MDAP. This is such a widely recognized and important concept that DoD has developed detailed Technology Readiness Assessment (TRA) levels and metrics and detailed guidance for specific acquisition milestones regarding the appropriate TRA level for identified critical technologies used in every MDAP.

Formal TRAs were originally developed by the National Aeronautics and Space Administration (NASA) in the 1970s and 1980s and, since then, NASA standards have served as the basis for the development of DoD TRA metrics,[20] as well as similar metrics for other agencies, such as the U.S. Department of Energy. Yet, while the development of TRA metrics was intended to standardize and greatly increase the precision of pre–MS B technology readiness assessments, this issue still remains a highly inaccurate science. While the TRA guidance provides a standardized framework for assessment metrics and includes measurable metrics, ultimately the final assessment of technology readiness and thus the gauging of the level of developmental risk at the beginning of a program still depends on the judgment, skill, and knowledge of a team of engineers who are trying to predict the risk involved in a potentially highly complex future development program. This is based, at least to some extent, on expert but nonetheless still subjective judgment, which can be influenced by many conscious and unconscious extraneous factors. One could argue that it is both a science and an art to accurately assess the readiness of key technologies and developmental risks at MS B for full-scale development using DoD TRA guidance. It is a difficult and demanding task that is not always done well.

An even more important and difficult complicating factor is the widespread recognition that, with the increased complexity of MDAPs and emergence of "systems of systems," there is a need to develop assessment metrics for measuring technology *integration* readiness and complexity as well as just technology readiness prior to MS B. The difficulty, risk, and readiness of integrating a variety of technologies into a single "system of systems" may be very challenging, although each individual technology needing integration may be accurately assessed as well known and relatively low risk with a relatively high TRA level. Thus, individual technology readiness and readiness for integration of multiple technologies in a new way are two completely separate and independent factors. The former is well known and has highly developed TRA metrics; the latter does not.

---

[20] Assistant Secretary of Defense for Research and Engineering, *Technology Readiness Assessment (TRA) Guidance*, Washington, D.C., U.S. Department of Defense, April 2011.

A good example of this problem can be found in the early developmental phases of the SBIRS satellite MDAP. Based on unpublished RAND research, we know that the Air Force carefully and extensively assessed the maturity of all the key component technologies being applied to the SBIRS space vehicle (SV) program and concluded that they were at appropriate readiness levels for MS B development but later experienced severe developmental challenges because of integration problems and complexities. This problem has now become widely recognized by acquisition analysts and engineers but is far from being solved by DoD.[21]

Unclear, unstable, or unrealistic requirements. There is a vast literature on the link between a "defective" requirements process and cost growth on major weapon system programs. Many studies of the MDAP acquisition process have noted that "establishing clear, affordable, and cost effective requirements has been a longstanding problem for the DoD acquisition process," and that "deficiencies in the process" have been "key factors in cost growth."[22] For decades, GAO and other independent expert observers have reported cost growth stemming from requirements "creep" and disconnects between the acquisition and user communities.[23] One example among many is the influential 2003 Young Report, which called out "undisciplined definition and uncontrolled growth in system requirements" as one of "five basic reasons" for significant cost growth and schedule delays.[24]

Our qualitative assessments of our six cases of extreme cost growth suggested that three key elements of the requirements process were major contributors to cost growth. The first was the lack of clarity and precision either in defining the desired system performance attributes or in translating those attributes into technical characteristics of the weapon system. This problem caused challenges in SBIRS and many other systems. Secondly, changing or unstable requirements after MS B also caused cost growth by necessitating excessive engineering change orders and redesigns. The RQ-4 Global Hawk was severely affected by this problem. And finally, the third element is closely related to lack of technological maturity, in that desired

---

[21] For DoD attempts to deal with this problem, see Brian J. Sauser and Jose Ramirez-Marquez, *System (of Systems) Acquisition Maturity Models and Management Tools*, Hoboken, N.J.: Stevens Institute of Technology, 2009; and Daniel Chien, *Ready or Not? Using Readiness Levels to Reduce Risk on the Path to Production*, Falls Church, Va.: General Dynamics, August 2011.

[22] Quotations from Gene Porter, Brian Gladstone, C. Vance Gordon, Nicholas Karvonides, R. Royce Kneece Jr., Jay Mandelbaum, and William D. O'Neil, *The Major Causes of Cost Growth in Defense Acquisition*, Vol. II, *Main Body*, Alexandria, Va.: Institute for Defense Analyses, IDA Paper P-4531, December 2009b, pp. 42–43.

[23] For a recent example, see GAO, *Defense Acquisition Process: Military Service Chiefs' Concerns Reflect Need to Better Define Requirements Before Programs Start*, Report to the Honorable James Inhofe, U.S. Senate, Washington, D.C., June 2015a.

[24] While the Young Report focused on national security space acquisition, its findings were broadly applicable to all MDAPs. Furthermore, four of our six cases of extreme cost growth were space programs, so its observations are particularly relevant. The full formal title of the Young Report was Office of the Under Secretary of Defense for Acquisition, Technology, and Logistics (USD[AT&L]), *Report of the Defense Science Board/Air Force Scientific Advisory Board Joint Task Force on Acquisition of National Security Space Programs*, Washington, D.C., May 2003.

system performance capabilities often are not realistically achievable within the funding and schedule constraints placed on the program, given the technical development necessary to achieve the desired performance capabilities. This was true for many of the sensors desired for the NPOESS weather satellite and that finally led to its cancellation.

Unrealistic cost estimates. This is an obvious and oft-mentioned cause of MDAP cost growth. A large literature exists on the organizational and behavioral factors that promote conscious or unconscious overoptimism in MS B cost estimates used for initial program baselines. As just one example, the extensive 2009 IDA study identified "erroneously low" initial MDAP cost estimates as "a major cause of acquisition cost growth."[25] The 2003 Young Report identified "unrealistic (cost) estimates" as another of its "five basic reasons for significant cost growth," concluding that "unrealistic projections of program cost . . . seriously distort management decisions and program content, increase risks to mission success, and virtually guarantee program delays."[26]

The Young Report argued that the overall acquisition system is heavily biased toward producing unrealistically low cost estimates. Our analyses of the MDAPs with extreme cost growth shows that several of the programs had access to much more realistic cost estimates than those actually selected as the formal baseline. This confirms the Young Report finding and suggests that overly optimistic cost estimates may be a serious problem that cannot be easily fixed.

Our second major category of characteristics of programs with extreme cost growth was suboptimal acquisition strategies and inappropriate program structure. Next, we focused on these two key sets of characteristics.

Acquisition strategies and program structures not tailored for level of risk. This characteristic is somewhat similar to the characteristic used in IDA's 2009 report labeled "excessive reliance on unproven management theories and acquisition strategies."[27] However, in our study, unlike the IDA study, this characteristic is almost always explicitly linked to an underestimation of technological or integration risk. Having underestimated the risk, programs were not structured to provide adequate oversight and workable mechanisms to manage risk effectively and respond appropriately to unanticipated developmental challenges. We also link this problem clearly to a penchant for acquisition strategies that envision a "single step to full capability" rather than an evolutionary or incremental approach.[28] As in the case of the IDA 2009

---

[25] Porter et al., 2009b, p. 37.

[26] Office of the USD(AT&L), 2003, p. 2.

[27] Porter et al., 2009b, p. 32.

[28] GAO and other critics of the MDAP acquisition process have published many reports over the years advocating multiple smaller steps toward an ultimate end-capability. Versions of this type of approach have variously been called "evolutionary acquisition," "incremental acquisition," "phased acquisition," "spiral development," and other similar nomenclature. For a recent account of this type of approach, see GAO, *Evolved Expendable Launch Vehicle:*

report, we also include in this category program structures and acquisition strategies that did not provide appropriate contractual or other incentives to the prime contractor or did not provide adequate guidance and oversight for the contractor.

Inappropriate use of MS B/C. This is a very specific type of acquisition program structure that was relatively common in the 1990s but is generally no longer used. Historically, it has been often linked to the problem of underestimating technological or integration complexity and risk or the level of RDT&E required in a program, while potentially leading to much greater program concurrency between the development and production phases. It was applied to programs perceived to be low in technological or integration risk or requiring little RDT&E. Typically, it was applied to programs that involved products conceived to be derivative from commercial technologies or existing systems, which are viewed essentially as nondevelopmental items. It was an attempt to reduce the regulatory burden and save time and cost by combining formal approval for the development stage with production approval. It was intended to be used when very little development was expected and production of an already-existing military or commercial item with little additional modification was anticipated. In nearly every case we are aware of, MDAPs that used this strategy and program structure ended up experiencing significant cost growth. This was usually because once preparations for production began, the procuring service and the contractor realized that much more substantial modification and development work was needed to meet the new service requirement. This often led to a stretch-out of the program and a delay in production or excessive concurrency with production beginning prior to the stabilization of a final production design.

A classic case of this problem is the T-6A Texan Joint Pilot Advanced Training System procured by the Air Force and the Navy. Originally intended to be a minimally modified off-the-shelf existing trainer called the Pilatus PC-9 (developed in Switzerland and produced under license in the United States), the basic aircraft eventually had to be extensively modified to meet U.S. service requirements, which led to substantial cost growth during the production phase.[29]

Of the two main categories discussed, the most significant, which spanned all six programs with extreme cost growth, was premature approval of MS B. Our research indicated that none of these programs was ready for MS B approval, usually for multiple reasons. Five programs were characterized by immature designs and technology or failure to recognize the complexity of system integration at MS B, combined with insufficient programmatic and technological risk-reduction efforts. These programs also suffered from unstable requirements that were incomplete, unclear, or disputed. Perhaps most striking, every one of the six programs suffered from serious cost-estimation issues. Nearly all the programs failed to place sufficient emphasis on the actual costs of similar or related predecessor programs. Five of the six programs passed

---

The Air Force Needs to Adopt an Incremental Approach to Future Acquisition Planning to Enable Incorporation of Lessons Learned, Report to Congressional Committees, Washington, D.C., GAO-15-623, August 2015b.

[29] See Mark A. Lorell and John C. Graser, *An Overview of Acquisition Reform Cost Savings Estimates*, Santa Monica, Calif.: RAND Corporation, MR-1329-AF, 2001.

MS B with at least some of the stakeholders or other interested parties aware that the MS B baseline cost estimates were unrealistic. In addition, most of the programs began with unclear, unstable, or unrealistic performance requirements and expectations and did not include an institutionalized process for modulating requirements in the interest of affordability as the program progressed. This issue is closely related to the category of immature technologies and integration complexity. The lack of realism in the cost estimates was also closely linked to the prior two elements, in that the difficulty and complexity of the required developmental and production efforts were underestimated, as were the effects of unrealistic or unstable requirements.

A second significant characteristic common across all six programs was the use of acquisition strategies or program structures not tailored for the actual levels of technical- and system-integration risk. None of the higher-risk programs systematically implemented evolutionary or incremental acquisition strategies or other overarching strategies specifically aimed at managing relatively high-technology and system-integration risks. Most of the programs were aware of *design components* or *specific technologies* that were high risk, and they

Table 1.2. Two Categories of Common Characteristics of Six MDAPs with Extreme Cost Growth

	AEHF	C-130 AMP	EELV	Global Hawk	NPOESS	SBIRS High
Premature MS B						
Immature technology; integration complexity	√	√		√	√	√
Unclear, unstable, or unrealistic requirements	√	√		√	√	√
Unrealistic cost estimates	√	√	√	√	√	√
Acquisition policy and program structure						
Acquisition strategy and program structure not tailored for level of risk	√	√	√	√	√	√
MS B/C	√		√	√	√	
PAUC growth (%)	95	193	273	152	154	279

SOURCE: Lorell, Leonard, and Doll, 2015.
NOTES: The bottom row shows PAUC growth for each program. There is little or no correlation between the number of characteristics evident in a specific program and the severity of that program's cost growth in percentage terms. Each MDAP is unique in context and circumstances, and this is why it is so important to convey the details of each case history, and ultimately to compare these six "worst of the worst" cases to best-performing MDAPs.

implemented effective technology risk-reduction measures for these specific areas. However, in many cases, the highest levels of risk resided in *overall system integration*, and we found no evidence of the implementation of comprehensive acquisition strategies for dealing with this type of risk. Several programs emphasized an acquisition approach employing a single step to full capability, which can increase risk. This may be acceptable if there is an urgent need for the capability, but acquisition and budget officials must be fully cognizant of the possible negative cost and risk implications.[30]

Other elements of the acquisition strategies adopted for these programs contributed to increased program risks. As discussed in our earlier companion report, most of these programs arose in the late 1990s during a period of new acquisition initiatives formulated by DoD that followed many of the acquisition elements developed under the Total System Program Responsibility (TSPR) concept first formulated in the early 1960s.[31] The revival of this TSPR-like approach in the 1990s stressed reduced government oversight, much greater contractor independence, and a strong emphasis on use of commercially derived technologies.[32] The commercial off-the-shelf (COTS) approach advocated vigorously by the Office of the Secretary of Defense (OSD) encouraged optimistic cost-savings estimates based on the more-extensive use of COTS technologies, civil-military integration (CMI), and commercial-type contracting and management approaches. This approach often led to an underestimation of the challenges of modifying and integrating commercially derived technologies and approaches into complex weapon system programs and lacked effective incentives to motivate the contractor to address problems early and effectively without direct government intervention.

Finally, approval of a combined MS B/C is risky unless the item to be procured is already adequately mature with a stable production design. Programs in the 1990s used simultaneous MS B and C approvals when it was thought that little or no RDT&E was needed. Yet the historical record suggests that, in these types of programs, the amount of RDT&E required to meet service requirements is often much more than anticipated, leading to higher costs and schedule delays. In addition, such programs appear to have a higher risk of suffering from concurrency of the development and production phases.[33]

---

[30] Some observers argue there is no evidence that the "single-step-to-full-capability" provides the end capability any faster than other approaches, and therefore should always be avoided.

[31] For further discussion on this topic, see our companion report, Lorell, Leonard, and Doll, 2015, pp. 7 and 38. Note the surveys of the six case studies in this earlier report.

[32] The focus on the use of COTS was a newer element added in the 1990s to the older TSPR concept.

[33] Recent RAND research suggests that acquisition phase concurrency, defined as the continuation of RDT&E activities beyond MS C, does not seem to contribute significantly to greater cost growth. The RAND research, however, did not distinguish between small overlaps of continuing RDT&E during the production phase—which are quite common on many MDAPs—and a nearly complete overlap of both the entire development and production phases, which is often the case with the use of the MS B/C approach. In-depth case studies of such programs as the RQ-4 Global Hawk provide ample evidence that, when there is little mission and design stability at MS B/C, and significant changes in design and configuration take place in the midst of series production, cost growth is inevitable. GAO has issued numerous reports on this topic. One example contrasting the overlapping acquisition phases of the Global Hawk program with the more structured evolutionary acquisition approach used on the MQ-1 Predator program can be found in GAO, *Unmanned Aircraft Systems: New DoD Systems Can Learn from Past Efforts to Craft Better and Less Risky Acquisition Strategies*, Report to the Committee on Armed Services, U.S. Senate, Washington, D.C., GAO-06-447, March 2006. See, also, GAO, *Unmanned Aerial Vehicles: Changes in Global Hawk's Acquisition Strategy Are Needed to Reduce Program Risks*, Report to the Chairman, Subcommittee on Tactical Air and Land Forces, Committee on Armed Services, House of Representatives, Washington, D.C., GAO-05-6, December 6, 2004.

As noted, to ensure that these characteristics were indeed the root cause of extreme cost growth of these six programs, it was necessary to compare and contrast them to key attributes of a similar number of best-performing recent Air Force MDAPs. The following subsection introduces those programs used to help verify our root cause analysis.

Four Programs with Low Cost Growth

To select the most-appropriate recent lower cost-growth Air Force MDAPs for comparison to the six MDAPs with extreme cost growth, we drew on essentially the same types of sources and general methodology described in the preceding section and in the companion 2014 document.[34] Based on a review of the roughly 30 relatively recent Air Force MDAPs from the SAR database, we were able to identify only four programs with clearly demonstrable low cost growth, adequate data, and other analytical requirements. These four programs were (in alphabetical order):

- C-5 Reliability Enhancement and Re-Engining Program (RERP)
- JDAM
- Small Diameter Bomb (SBD) Increment I (SDB I)
- WGS Block I.

It was important for our analysis to carefully select better-performing programs that were truly representative of the same technology era and acquisition environment as the six extreme cost-growth programs. To ensure a more-consistent and fair comparison, the selection criteria for these programs were slightly more complex than for the programs with extreme cost growth and therefore required further explanation. First, we sought Air Force MDAPs that originated during the same relatively narrow time frame to avoid possible complications from comparing systems from differing acquisition policy environments or technology eras. This is particularly true for programs originating in the 1990s, which was a unique period of wide-ranging acquisition reform efforts and experimentation. Thus, the four lower cost-growth systems we selected passed their MS B decision, or beginning of full-scale development, between 1995 and 2003. In the case of the six programs with extreme cost growth, their MS B or combined MS B/C decisions took place during virtually the same time frame, from 1994 to 2003, thus beginning development in the same acquisition policy and technology environment.

Another important consideration, particularly for our sponsor, was that the programs chosen should still be part of the active acquisition portfolio to facilitate interaction with program offices, and that they have cost histories of adequate length to provide a reasonable amount of cost data for analysis. Thus, such well-performing programs as LGM-30 Minuteman II Guidance Replacement Program and Minute III Propulsion Replacement Program were excluded because they were no longer active reporting elements of the acquisition portfolio when we conducted

---

[34] Leonard and Wallace, 2014.

our analysis. While the AIM-120 Advanced Medium Range Air-to-Air Missile was still being procured, it went through its original MS B many years prior to our target period of the mid-1990s through early 2000s, so it too was excluded.[35] Recent programs with insufficient cost data to demonstrate ultimate cost growth were also excluded, such as the B-2 Extremely High Frequency SATCOM and Computer Increment 1, which went through MS B in 2007.

It might surprise some readers to learn that two of the four lower cost-growth systems we selected, the C-5 RERP and the overall WGS system—including Block I, Block II, and Block II Follow-On (B2FO), as opposed to Block I alone—both experienced Nunn-McCurdy cost breaches.[36] However, while a Nunn-McCurdy breach is an official indication of significant cost growth, it is not necessarily a sign of unusually high cost growth. Cost-related Nunn-McCurdy breaches can occur with an increase of as little as 15 percent (this is what happened in the case of the C-5 RERP program).[37] Prior RAND research shows that the *average* MDAP, which delivers at least 25 percent of the units originally planned at MS B, experiences 60-percent cost growth from MS B. In this context, a program that suffers only a 15-percent increase in cost, thus triggering a formal Nunn-McCurdy breach, may still end up being reasonably classified as a lower cost-growth program. Also, as explained in greater detail later in this report, the RAND cost methodology, which normalizes costs at the end of the program for quantities planned as MS B and focuses on PAUC, which includes both overall development and quantity-adjusted total procurement costs, shows that the originally planned WGS Block I program experienced little cost growth, although the overall program—including Blocks I, II, and B2FO—experienced substantial cost growth.

Thus, given these clarifications and qualifications regarding the selection process, only the four programs listed above met all our criteria as lower cost-growth Air Force MDAPs appropriate for comparison with our earlier list of Air Force MDAPs with extreme cost growth.

Table 1.3 presents the same five categories of cost growth using the same metrics and end point as shown for the six extreme cost-growth MDAPs shown in Table 1.1. In stark contrast to

[35] At the time this research was conducted in 2014, SDB I was no longer reporting SARs but was still in production. A closely related follow-on system, the SDB II, was still reporting SARs.

[36] Nunn McCurdy refers to U.S, Code, Title 10, Section 2433, Unit Cost Reports (UCRs), January 7, 2011. This amendment sponsored by Senator Sam Nunn and Congressman David McCurdy was inserted in Title 10 of the National Defense Authorization Act of FY 1982. It establishes a series of cost-growth and schedule thresholds for all MDAPs that, if breached, require formal reporting by program managers to Congress and may require the Secretary of Defense to formally certify to Congress the need for a program for it to continue.

[37] There are two levels or categories of statutory breaches established by the Nunn-McCurdy legislation: significant and critical. The former has much less serious consequences than the latter. Each of the two levels or categories of breaches has two threshold cost-growth percentages: one for comparison to the current acquisition program baseline (APB) estimate, and one for the original APB estimate. The two threshold cost-growth levels for significant breaches are 15 and 30 percent of the average procurement unit cost (APUC) or PAUC. The two threshold cost-growth levels for critical breaches are 25 percent and 50 percent, respectively. See Department of Defense Instruction, 2015.

the six programs with extreme cost growth, these four programs show virtually no categories with extreme cost growth, with the notable exceptions of JDAM and WGS. The three extreme

Table 1.3. Four Better-Performing Cost-Growth Air Force MDAPs

Program	MS B or B/C	Budgetary Cost Growth (%)			FY 2015 M$ Growth	Unit Cost Growth (%)	
		Development	Procurement	Program		Procurement	Program
C-5 RERP	December 2001	2	−39	−32	−3,600	22	18
JDAM	October 1995	17	***153***	127	4,500	−19	−12
SDB I	October 2003	−5	−49	−39	−800	−20	−26
WGS	January 2001	113	***249***	***224***	2,800	−14	9

NOTES: Percentages shown in bold italics represent extreme cost growth, defined as cost growth more than one standard deviation above the mean for that measure. For WGS, budgetary cost growth includes all blocks (Block I, Block II, B2FO). Unit cost growth has been adjusted to baseline quantities (Block I only).

cost-growth numbers shown in two budgetary cost-growth categories for these two programs are in part an artifact of the cost-growth assessment methodology and do not represent extreme cost growth in accordance with RAND methodology. The problem with budgetary cost-growth metrics is that they do not make adjustments for changes in quantities. Sometimes, budgetary cost growth is caused by great acquisition success, leading the services to demand far more of the system than originally anticipated, thus increasing budgets. This is what happened in the case of JDAM. Note that, in all other cases, particularly in the most-important categories for our comparisons (the procurement and program-acquisition unit cost-growth categories, where adjustments for quantity changes are made), all four programs show at worst modest cost growth. Indeed, out of the eight cost-growth numbers in these two categories, only three are more than zero. The remaining six numbers show significant *negative* cost growth in the cases of JDAM, SDB I, and WGS. More discussion of these results is in Chapter Three.[38]

[38] In the case of JDAM, extreme cost growth in the budgetary program cost-growth column is due to the government's authorization of a dramatic increase in JDAM production quantities to use in the war on terrorism after the September 11 attacks. The large budgetary increase required to fund this wartime surge in JDAM kits exceeded the original budget plan and thus technically appears as significant cost growth. However, the average production unit cost for JDAM stayed the same, so this does not represent true cost growth. In the case of WGS, where budgetary procurement and program cost growth are shown as extreme, some of the cost growth is real, but does not apply to the original first part of the program we examined as laid out at MS B. We examined only the first part of the WGS program encompassing the Block 1 satellites, not the Block II or B2FO satellites, because only the Block I satellites had been envisioned at MS B. So, the extreme budgetary cost growth shown or WGS is also an artifact of the cost-estimating methodology and is not an accurate representation of PAUC, which took place based on the original plan as laid out during the original MS B. Of course, this situation can play both ways. The large budgetary procurement and program cost declines shown for the C-5 RERP are also partly an artifact of this cost methodology. Here, the true reason that the cost declined was because the Air Force decided, because of a variety of complex factors, to significantly cut the planned quantities of C-5 RERPs. This was because some estimates envisioned large future increases in unit costs. But, in addition, it appears that the cost-growth projections that led to

Table 1.4 presents some basic programmatic information for these four programs regarding the MS B date, overall program cost, RDT&E cost, and PAUC growth as of the December 2013 SARs. These factors can be compared with the same factors for the six cases with extreme cost growth shown in Table 1.5. Even without reviewing any details of the case histories of these programs, several comparisons are immediately evident. First, as noted earlier, they all passed through MS B during the same seven-year period, from 1995 through 2003. Indeed, one-half of all the programs in each group passed through MS B in 2001. Thus, they were all subject to roughly the same acquisition policy and regulatory environment in that regard.[39]

The extreme cost-growth group includes three space systems and a space launch system, an aircraft modification program, and a large high-altitude remotely piloted vehicle (RPV). The low cost-growth group is made up of two high-volume smart munitions, an aircraft-modification

Table 1.4. Cost Overview of Four Air Force MDAPs With No or Low Cost Growth

Program	MS B	Total Cost (FY 2015 $B)	RDT&E COST (FY 2015 $B)	PAUC CG (%)
C-5 RERP	December 2001	7.5	1.9	18
JDAM	September 1995	8.2	0.8	–12
SDB I	October 2003	1.2	0.4	–16
WGS Block I	January 2001	4.1	0.5	9

NOTE: Cost growth is shown for PAUC. The PAUC growth shown for WGS covers only the first three SVs, which made up Block I. For further explanation, see Chapters Two and Three. Data are from the December 2013 SARs. Costs are in FY 2015 dollars. CG = cost growth.

Table 1.5. Cost Overview of Six Air Force MDAPs with Extreme Cost Growth

Program	MS B	Total Cost (FY 2015 $B)	RDT&E Cost (FY 2015 $B)	PAUC CG (%)
AEHF	November 2001	14.3	8.5	95
C-130 AMP*	July 2001	6.6	2.1	193
EELV	October 1998	61.2	2.3	273
RQ-4	March 2001	10.1	4.3	152
NPOESS*	August 2002	13.2	7.7	154
SBIRS High	November 1996	19.5	11.9	279

NOTES: Cost growth is shown for PAUC. The C-130 AMP and the NPOESS satellite were both canceled before completion of development because of high cost growth. Data are from the December 2013 SARs. Costs are in FY 2015 dollars. CG = cost growth.
* Programs were terminated or truncated in the FY 2013 President's Budget.

---

the Nunn-McCurdy breaches were somewhat inflated. Also, the Air Force and the contractor implemented stringent and effective cost-reduction measures on the remaining aircraft. Therefore, reduction in procurement numbers was not the main cause for C-5 RERP reduction in expected cost growth. As shown in Table 1.3, APUC and PAUC growth remained relatively low for the C-5 RERP program, and APUC and PAUC are normalized for changes in quantity, thus showing that reduction in procurement was not the main cause for lack of high cost growth. The details and justification of the RAND treatment of WGS, JDAM, SDB I, and C- 5 RERP cost growth are all extensively reviewed in Chapter Two.

[39] Three different versions of DoDI 5000.2 came out during this period: in 1996, 2000, and 2003. However, most of the MDAPs during this period were influenced by the same major acquisition reform legislation coming out of Congress (such as the Federal Acquisition Streamlining Act of 1994 and the Federal Acquisition Reform Act of 1996), as well as the Clinton administration acquisition reform policies.

program, and a space system. The six extreme cost-growth programs tend to be larger in dollar value than the low cost-growth programs. Of course, this is due in part to the extreme cost growth experienced by the large dollar-value programs. Chapter Three examines this issue in greater detail.

To gain better insights into the low cost-growth programs and whether they support the hypothesis that the key characteristics of the programs with extreme cost growth were indeed the key cost drivers and root causes of cost growth, it is necessary to examine the case histories of the low cost-growth programs and compare them carefully to the case histories of the extreme cost-growth programs. The core questions we sought to answer with our case studies of the four programs with low cost growth were as follows:

- Did these programs possess similar characteristics and experience similar challenges as the extreme cost-growth programs? Why or why not?
- If so, how were they managed successfully?
- What else is different about these programs?

Chapter Two provides overviews of each of the four programs with low cost growth and discusses their key attributes organized under the five main subcategories of key characteristics as found in the programs with extreme cost growth. Chapter Three also provides our overall assessment of these four programs: whether they tend to confirm the hypothesis stated above and whether they reinforce our earlier report's identification of factors mitigating extreme cost growth. The four low cost-growth programs are reviewed in alphabetical order.

2. Case Studies

C-5 Reliability Enhancement and Re-Engining Program

Figure 2.1. C-5M Super Galaxy

SOURCE: At Dover Air Force Base, Delaware. Photo courtesy of Lockheed Martin.

Summary Overview

With Lockheed Martin as the prime contractor, RERP is the second of phase of a two-phase modernization program for the C-5, the Air Force's largest air lifter. It is an aging aircraft, with the first examples procured in the late 1960s. Phase I, AMP, is the baseline for the RERP.[40] This initial modernization phase primarily equips the C-5 with digital avionics, which enables it to fly in civil airspace by the most-direct routes, at the most-advantageous altitudes, and with the most-efficient fuel usage and cargo loads. This is followed by the RERP, which is Phase II. The RERP primarily replaces the current General Electric (GE) TF39 engine with a more-reliable, COTS GE F138-GE-100 turbofan engine, a close variant of the commercial GE CF6 engine.[41] In addition, approximately 70 other C-5 subcomponents and parts are upgraded or replaced. The

---

[40] The Phase I C-5 AMP program is technically a separate program from C-5 RERP. But planners assumed AMP would be concluded prior to initiation of C-5 RERP modifications.

[41] DoD, *Selected Acquisition Reports (SAR): C-5 Reliability Enhancement and Re-engining Program (C-5 RERP)*, Washington, D.C., December 2001b–2014b, p. 6.

powerful new engines deliver 22 percent more takeoff thrust, achieve 30-percent shorter takeoff distances and enable 58-percent faster time-to-climb to cruising altitude. After completing both phases of the modernization program, C-5s are designated the C-5M Galaxy aircraft. Figure 2.2 shows the major events in the C-5 RERP acquisition time line.

Figure 2.2. Time Line of Major C-5 RERP Production Events

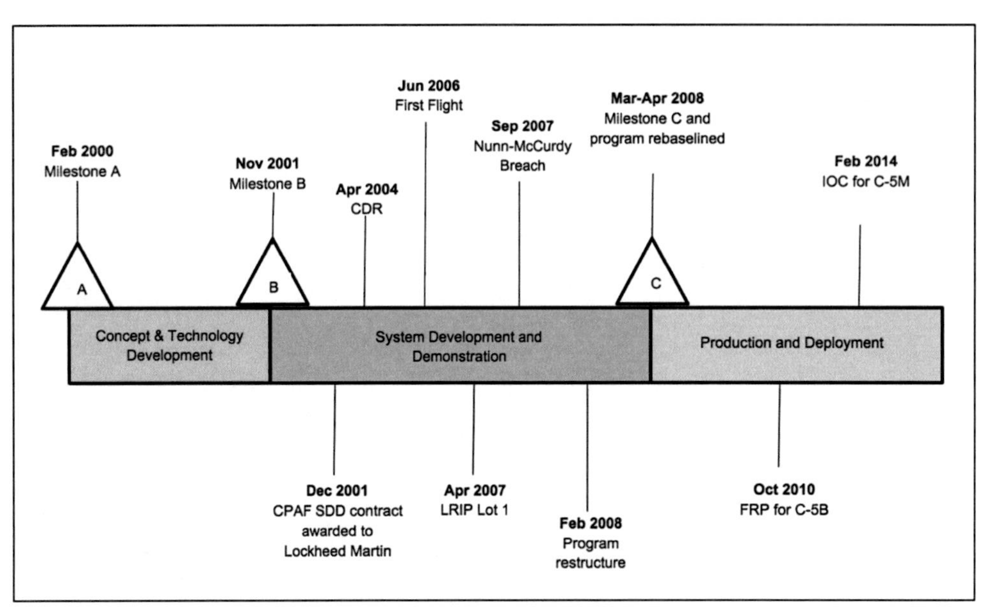

SOURCE: RAND graphic assembled from annual C-5 RERP SAR data.
NOTE: CDR = critical design review; IOC = initial operational capability;
CPAF = cost-plus-award fee; SDD = system development and demonstration;
LRIP = low-rate initial production FRP = full-rate production.

During development, the C-5 RERP program experienced funding, schedule, and technical issues, which ultimately led to a Nunn-McCurdy breach in 2007 and a program restructuring.[42] The Air Force had released a high-profile Mobility Capabilities Study in 2005, which determined that a total of 292 heavy lifters (180 C-17s and 112 upgraded C-5s, including three SDD aircraft)[43] were needed to meet the projected future lift requirement of 33.95 million ton-miles per day.[44] The planned C-17 numbers remained stable, but, by late 2006 and early 2007, skepticism mounted regarding the ability to meet the C-5 production goals because of growing cost and technical issues.[45] In early September 2007, a new service-cost position issued by the Air Force projected more than 50-percent cost growth if the current C-5 RERP program continued with its plan for 112 C-5Ms. The newly calculated APUC was much higher than the

[42] "Saving the Galaxy: The C-5 AMP/RERP Program," *Defense Industry Daily*, June 18, 2015.

[43] This quantity of 180 C-17 was later extended by ten by FY 2007 bridge supplemental, resulting in a planned 190 C-17 aircraft. Eventually, a total of 213 were procured, ending in September 2013.

[44] "USAF Directed to Find $1.8 Billion for C-5M, Consider Logistics Support," *Inside the Air Force*, May 5, 2008.

[45] "Senators Press Air Force Brass on Airlift Recapitalization Needs," *Inside the Air Force*, March 23, 2007.

original unit cost that was reported in November 2001. Air Force acquisition officials cited AMP schedule slippage and other unanticipated problems as the root cause for the projected cost growth. For example, older C-5As that were slated to be "RERP-ed" along with C-5Bs were found to have unanticipated age-related problems, such as corrosion and cracks.[46] More specifically, the Air Force attributed the projected increases in RERP costs to additional costs for touch labor and supplier parts.[47] In addition, unanticipated redesign of the engine pylons and strengthening of the wing structure were deemed necessary following work on the prototype aircraft.

Through 2005, DoD expected to spend a total of about $10 billion (2000 base-year [BY]$) for 112 C-5 RERP aircraft.[48] In 2007, a senior Air Force official testified to Congress that the C-5 RERP and AMP programs together had been originally estimated to cost $12 billion for 112 C-5Ms, with most of the cost attributed to the RERP.[49] The Air Force's revised estimate in 2007 for the RERP upgrade envisioned a cost increase to $17.5 billion. This was about $4.5 billion more than Lockheed Martin's own 2007 revised estimate of $13.3 billion at that time, which was rejected by the Air Force as unrealistically optimistic. The Air Force stood by its revised higher estimate and, as a result, officially declared a Nunn-McCurdy breach to Congress on September 27, 2007.[50]

The Office of the USD(AT&L) restructured the program in February 2008 by limiting the RERP upgrade to just 52 C-5Bs (49 production aircraft and three SDD aircraft). This restructuring removed the 60 older C-5A aircraft from the RERP, although they stayed in the

---

[46] At the time of the September 27, 2007, Senate hearing that discussed the Nunn-McCurdy breach, C-5As were approximately 29–40 years old, compared with the much-younger C-5Bs. Some estimates describe C-5As as being approximately 40 years old. "Statement of Sue C. Payton, Assistant Secretary of the Air Force for Acquisition, U.S. Air Force, Accompanied by Diane M. Wright, Deputy Program Executive Officer for Aircraft, Aeronautical Systems Center, Wright-Patterson Air Force Base," in *Cost Effective Airlift in the 21st Century*, hearing before the Federal Financial Management, Government Information, Federal Services, and International Security Subcommittee of the Committee on Homeland Security and Governmental Affairs, U.S. Senate, 110th Cong., 1st Sess., Senate Hearing 110–410, September 27, 2007, pp. 8–10 and 49–57.

[47] "Statement of Sue C. Payton . . . ," 2007.

[48] DoD, 2005b.

[49] A 2010 CRS report describes this $12-billion estimate as comprising approximately $900 million for the AMP program and $11.1 billion for the RERP program. See U.S. Government Accountability Office *Defense Acquisitions: Timely and Accurate Estimates of Costs and Requirements Are Needed to Define Optimal Future Strategic Airlift Mix*, Washington, D.C., Report to the Subcommittee on Air and Land Forces, Committee on Armed Services, House of Representatives, GAO-09-50, November 2008, pp. 8–9 (cited in Jeremiah Gertler, *Air Force C-17 Aircraft Procurement: Background and Issues for Congress*, Washington, D.C.: Congressional Research Service, 7-5700, RS2273, December 22, 2009, p. 11, footnote 24; and William Knight and Christopher Bolkom, *Strategic Airlift Modernization: Analysis of C-5Modernization and C-17 Acquisition Issues*, Washington, D.C.: Congressional Research Service, RL34264, October 22, 2008, p. 8.

[50] Other considerations were also in play. The Air Force was facing real prospects of serious budget cuts and was concerned about funding the new KC-X aerial tanker acquisition program. In addition, Congress wanted the Air Force to procure more C-17s than the Air Force thought necessary, which placed greater budgetary pressure on the Air Force while providing more airlift capacity by increasing the size of the future C-17 fleet. See "Statement of Sue C. Payton . . .," 2007.

AMP program. Despite lower procurement numbers, this reduced projected growth in unit costs significantly because the older C-5As were the ones with the major "over-and-above" reliability issues and age-related parts problems when they were opened up.

After the program restructure, with the highly problematic A-models removed from the RERP, all 52 B-models were slated to be the first to receive the AMP upgrade, followed by the RERP upgrade.[51] The new program evolved rapidly as the following events took place from March to June 2008:

1. The Defense Acquisition Board conducted a successful Milestone C review.
2. The RERP was re-baselined following a Nunn-McCurdy recertification.
3. The RERP was approved for MS C.[52]

The quantity cut and restructuring of the program improved the unit cost expectations and substantially reduced overall projected program costs, leaving more funds available for such higher-priority programs as the KC-X aerial tanker. The high cost growth that was projected prior to the breach was replaced by expectations of little to no cost growth following MS C in 2008. These expectations proved largely valid. Cost growth from the original MS B through the December 2013 SAR for procurement, with adjustments for quantity changes, was a relatively moderate 22 percent and cost growth for RDT&E only 2 percent.[53] This demonstrates that the Air Force and the contractors' intensive efforts to control and reduce unit costs after the program restructure had a significant effect, since these numbers are adjusted for original quantities.

Given this relatively modest cost growth, how do the characteristics of the C-5 RERP compare with similar key characteristics identified by our earlier report on the six programs with extreme cost growth?

C-5 RERP Program Characteristics

Technology Maturity and Integration Complexity

Technological problems were common in the earlier AMP phase of the program but less so during the RERP. Delays in the AMP program played a significant role in slowing the C-5 RERP effort and contributed to RERP cost growth and the Nunn-McCurdy breach in 2007. For example, the AMP upgrade suffered from unstable software systems and system-integration challenges. In fact, operational testing and evaluation underwent a one-month suspension because of avionics software immaturity and maintenance technical order issues. These factors combined with others to cause AMP schedule delays and cost overruns, which drove early cost

---

[51] Most, but not all, of the C-5As remained in the C-5 AMP program.

[52] Gertler, 2009.

[53] Data derived from RAND cost analysis.

growth on RERP.[54]

The period following the 2008 restructuring saw few technical problems for the RERP. RERP modernization primarily entailed engine upgrades, and the GE commercial-derived F138-GE-100 turbofan engine was a mature and well-tested system.[55] The GAO, in its 2009 and 2010 *Assessments of Selected Weapons Programs*, stated the "RERP critical technologies are mature and its design is stable."[56] It also described the RERP as continuing to buy and integrate commercially available items,[57] in addition to the F138-GE-100 engine. Other documents that discuss key aspects of the C-5 RERP acquisition strategy emphasize the reliance and integration of commercial technology. GAO stated, "By virtue of the fact the RERP integrates commercial technology and COTS items used by other military and commercial aircraft, there is no need to evaluate its technology and production maturity."[58]

Requirements Realism and Stability

Prior to the program restructure in 2008, quantity requirements for the mix of C-5Ms versus C-17s were not fully established. A 2010 CRS report that discussed C-17 and C-5 aircraft procurement noted that DoD's strategic airlift requirements with respect to the C-17 and C-5 had fluctuated substantially and "evolved over the years."[59] As noted, the 2005 Mobility Capabilities Study, in which DoD identified a minimum requirement for 292 strategic airlift aircraft, assumed procurement of 112 of the fully upgraded C-5Ms. The combination of development and remanufacturing problems, especially with the C-5As, delays on the C-5 AMP program, all led to the Nunn-McCurdy breach in 2007 and restructuring of the program. The increase in planned Air Force procurement of C-17s, combined with the lower cost of the C-5 RERP program after restructuring, meant that the overall airlift requirement, as well as other Air Force priorities could

---

[54] "Statement of Sue C. Payton . . . ," 2007.

[55] DoD, 2001b–2014b, p. 6. The F138-GE-100 engine is the military derivative of the commercial GE CF6 engine powering many commercial wide-body passenger transports.

[56] U.S. Government Accountability Office, *Defense Acquisitions: Assessments of Selected Weapons Programs*, GAO-10-388SP, March 2010a.

[57] GAO, 2010a.

[58] GAO, 2010a. Unlike some development programs, such as SBIRS, EELV, C-130 AMP, C-5 AMP, and even WGS Blocks II and Block II Follow On, that found incorporation of commercial-derived technology much more challenging than expected, the C-5 RERP clearly benefited from a COTS approach. As noted elsewhere, the most expensive single component of the RERP program was the engine, which was a little changed, widely used, and very mature COTS transport aircraft engine. The other programs used commercially derived technologies or major components that were not fully matured or fully applicable to military systems without significant modification. Nonetheless, the RERP program still encountered more issues than anticipated in integrating the new COTS engine into the C-5 airframe.

[59] Gertler, 2009.

still be met with the reduced numbers of C-5Ms in the restructured program.

After 2008, requirements stabilized, and the program continued to move ahead successfully. In 2008, the Joint Requirements Oversight Council (JROC) conducted a C-5 RERP assessment that concluded the C-5 AMP and RERP programs were essential to national security. The council specified that modernizing 52 C-5 aircraft best met the mission essential criteria set forth by the JROC. This further stabilized the RERP program.

By 2010, RERP was considered technologically mature and stable, and all seven lots of the restructured program were successfully exercised.[60]

Cost-Estimate Realism

Cost estimates for C-5 RERP proved to be unrealistic prior to the 2007 breach. Several issues led to an increase in the program cost estimates. MS B cost-estimating issues are also delineated in a different document that explains the Nunn-McCurdy certification process. Therein, a Cost Analysis Improvement Group assessment ascribes the drivers of high APUC growth that occurred from 2001 to 2007 (estimated at 52.7 percent above the 2001 baseline) to the following factors: [61]

- "**Material Cost Growth**. Material costs included in the development proposal at the onset of the program were significantly lower than those estimated at Milestone B many years later. The higher material costs at Milestone B reflected price escalation for certain raw materials, especially strategic or specialty metals. This accounted for 18.2 percentage points of the estimated 52.7% APUC growth."
- *"**Estimation**. Spares to support initial deployment were underestimated, in addition to government-furnished equipment and mission support. This accounted for 16.5% percentage points of the estimated 52.7% APUC growth."*[62]
- "**Labor Cost Growth**. This resulted from (1) an increase in hours required to perform installation of prime mission equipment, in addition to "over and above" repairs to modernize the three SDD aircraft, and (2) a significant increase in labor rates that were reflected in the prime contractor's Forward Pricing Rate Agreement. This accounted for 12.3 percentage points of the estimated 52.7% APUC growth."
- "**Production Rate**. The annual production rate quantities for the baseline program assumed an economic order quantity of 12 aircraft per year. In 2007, the Air Force felt unable to commit to this rate of production throughout the planned procurement due to

---

[60] DoD, 2001–2014a, pp. 31–34.

[61] Quoted from U.S. Senate, 110th Cong., 1st Sess., Cost Effective Airlift in the 21st Century: Hearing before the Federal Financial Management, Government Information, Federal Services, and International Security Subcommittee of the Committee on Homeland Security and Government Affairs, Washington, D.C.: U.S. Government Printing Office, Hearing 110-410, September 27, 2007.

[62] Italicized by the authors for emphasis.

budget shortfalls and changing budget priorities, and sought a much lower initial annual production rate. This accounted for 5.7 percentage points of the 52.7% APUC growth."[63]

Compared with the six programs with extreme cost growth, however, these cost-estimation issues were relatively modest. And unlike the extreme cost-growth programs, many of the factors involved were out of the control of the program office; these included the delays and cost growth on C-5 AMP, the budget cuts, Congress's insistence on funding more C-17s than the Air Force thought it needed, and the changing Air Force budget priorities, especially the need to fund the KC-X program. As noted, we calculate that, through the December 2013 SAR, APUCs for the entire RERP since MS B grew only a relatively modest 22 percent, with RDT&E cost growth at just 2 percent. Of course, the projected cost growth in both RDT&E and procurement, which led to the Nunn-McCurdy breach in 2007, was also not realized because the 60 older and more-problematic C-5As, which were expected to be the main cause of the cost growth, were removed from the program.

Acquisition Strategy and Program Structure

Prior to the 2007 Nunn-McCurdy breach and subsequent to the re-baselining in 2008, the acquisition strategy entailed modernizing 112 C-5 aircraft by equipping them with both the AMP and RERP upgrades, after which they would be designated C-5M Galaxies. The Air Force planned to upgrade the B models first, but the Nunn-McCurdy breach exposed ongoing schedule slippage and mounting cost growth.

After the re-baselining, the acquisition strategy shifted to focus on four items:

- The quantity was cut to 52 aircraft, which increased the program's top-line affordability.
- The high-risk factors driving the production cost increases projected in 2007 were eliminated. Prior to the breach, Air Force officials and other stakeholders frequently cited the difficulty of modernizing C-5A aircraft, which were 29 to 40 years old and beset with structural issues. By comparison, applying the RERP upgrade to the newer B-models proved easier and more cost effective.[64]
- Similar to other programs mentioned throughout this chapter, Lockheed Martin used proven COTS items, most importantly, GE's F138-GE-100 engine, which was similar to the commercial CF6, of which each C-5M required four.
- The C-5 system program office (SPO) and Lockheed Martin agreed to a series of fixed-price contracts with economic price adjustment (FPEPA) for FRP lots 1 through 7.[65] Similar to firm fixed-price contracts, FPEPA contracts hold the contractor to the agreed-

---

[63] Quoted from U.S. Senate, 2007.

[64] "Statement of Sue C. Payton . . . ," 2007; and "USAF Directed to Find $1.8 Billion for C-5M, Consider Logistics Support," 2008.

[65] DoD, 2010b, provides contract information on lots 1 through 3 and discusses the FPEPA type of contract; DoD, 2014b, provides identical information for lots 4 through 7.

upon unit lot price for the aircraft, with possible adjustments for inflation and other economic factors.

Milestone B/C

The C-5 program avoided excessive overlap of RDT&E and production phases, unlike most of the programs that have experienced extreme cost growth. Indeed, the C-5 RERP can be characterized as a phased- or incremental-acquisition program. The AMP effort was split out and had to be completed before RERP production began. Production was delayed when the cost estimates rose substantially in 2007 (in part because of the development of problems and delays in the AMP phase, as well as problems with the C-5A conversions), leading to the Nunn-McCurdy breach and program restructure. MS C, the FRP authorization, was approved in accordance with the original program schedule in March 2008. But FRP was spun up more slowly than originally envisioned. Actual FRP of C-5B RERPs was not launched until October 2010, almost two years after the originally planned 2001 objective date. Seven months prior to the beginning of FRP, fully dedicated Qualification Operational Test and Evaluation, including the final report of the Air Force Operational Test and Evaluation Center, was completed. Thus, there was virtually no overlap of the development, test, and production phases. The C-5B RERP was fully mature and ready for production when FRP began.

C-5 RERP: Summary of Findings on Root Causes of Low Cost Growth

Despite developmental problems with the C-5 AMP program and with modifications to the C-5As, the Air Force was able to restructure the C-5 RERP in a manner that better fit budgetary realities and priorities, helping it to achieve relatively low PAUC growth (adjusted for quantity changes) compared with the original MS B estimates, even with a substantial cut back in total procurement numbers:

- The C-5 RERP program was part of an overall incremental upgrade program (AMP plus RERP), which heavily focused on use of COTS and fixed-price production contracts to reduce and control costs. The new engine, the single most-expensive element of the RERP, was a reliable COTS engine from GE.
- Because of unanticipated technical problems largely associated with the older C-5As, budget cutbacks, delays in the C-5 AMP program, congressional pressure to buy more C-17s than planned, and changing Air Force budgetary and procurement priorities, the programs was restructured in 2007 to focus solely on the C-5Bs.
- Despite the challenges facing the program, the Air Force carefully and judiciously restructured the program, ensuring readiness for production at MS C, completing all qualification operational testing prior to FRP, working closely with the prime contractor to keep costs down, and using FPEPA contracts to control costs.

The C-5 RERP effort successfully entered MS C following the 2007 breach and 2008 restructuring. It achieved a relatively low 22-percent APUC growth (adjusted for quantity changes) and only 2-percent cost growth for RDT&E from the original MS B through the period up to the December 2013 SAR.[66]

Table 2.1 assesses the C-5 RERP in terms of the key characteristics driving cost growth on the six programs with extreme cost growth. Notice that C-5 RERP evinces only two of these characteristics in the first category and only to a relatively moderate extent. C-5 RERP shows the first characteristic because program officials and the contractor significantly underestimated the complexity and difficulty of implementing the planned airframe upgrades to the aircraft, especially to the older C-5As. In part because they underestimated the technical challenge, they also developed unrealistically optimistic cost estimates. Nonetheless, compared with the six MDAPs with extreme cost growth, the errors committed on C-5 RERP were relatively modest.

Table 2.1. C-5 RERP Had Low Cost Growth Compared with Most MDAPs, but the Highest Cost Growth Among the MDAPs with Low Cost Growth

	C-5 RERP
Premature MS B	
Immature technology; integration complexity	√
Unclear, unstable, or unrealistic requirements	
Unrealistic cost estimates	√
Acquisition policy and program structure	
Acquisition strategy and program structure not tailored for level of risk	
MS B/C	
PAUC growth (%)	18

In addition, the other characteristics common to the high cost-growth programs are not found. C-5 RERP requirements (except for total quantity) remained remarkably stable. The acquisition strategy and program management proved to be highly responsive and flexible in the face of unanticipated challenges. The program was rapidly and successfully restructured after the Nunn-McCurdy breach. Program officials quickly identified the problem as the older C-5As which, when opened up, had far more problems and areas requiring attention than the newer C-5Bs. The success of this approach is demonstrated by the fact that, even when the final cost is normalized to the original MS B quantities, PAUC growth only rose 18 percent, significantly below the average cost growth of all MDAPs, and way below the originally projected Nunn-McCurdy breach based on the 2007 cost projection. Boeing and the SPO worked very closely to control costs after the program restructure, and it appears they have done a reasonable good job.

[66] Data derived from RAND cost analysts.

Joint Direct Attack Munition

Figure 2.3. GBU-31 JDAM Tail Kit

SOURCE: Photo taken at a forward deployed location in Iraq; courtesy of Staff Sgt. Jessica Kochman.

Summary Overview

Boeing's JDAM has been described as "the linchpin" of DoD's acquisition streamlining activities in the 1990s.[67] JDAM was designated an ACAT ID program in 1991.[68] The reform environment of the early 1990s shaped JDAM's cost trajectory and its acquisition strategy. The 1994 Federal Acquisition Streamlining Act of 1994, the Federal Acquisition Reform Act of 1996, and the Clinton administration acquisition-reform initiatives all combined to encourage DoD and the Air Force to streamline the program's acquisition process.

The JDAM program sought to quickly develop a low-cost, low-risk "smart" munitions tail kit using an inertial navigation system (INS) and the Global Positioning System (GPS) to accurately acquire and hit a target at long range and under a wide range of adverse weather and

---

[67] Lt Gen General George Muellner, former Principal Deputy Assistant Secretary of the Air Force for Acquisition, characterized JDAM as such in 1996. See Mark A. Lorell, Julia F. Lowell, Michael Kennedy, and Hugh P. Levaux, *Cheaper, Faster, Better? Commercial Approaches to Weapons Acquisitions*, Santa Monica, Calif.: RAND Corporation, MR-1147-AF, 2000, p. 139.

[68] An ACAT ID is an MDAP with a "Dollar value for all increments of the program estimated by the Defense Acquisition Executive (DAE) to require an eventual total expenditure for research, development, and test and evaluation (RDT&E) of more than $480 million in Fiscal Year (FY) 2014 constant dollars or, for procurement, of more than $2.79 billion in FY 2014 constant dollars." The Milestone Decision Authority for an ACAT ID is the DAE, that is, the USD(AT&L). See DoDI 5000.02, 2015.

environmental conditions. JDAM fulfilled this need by providing a "strap-on" guidance tail kit that could be attached to standard 1,000-pound MK-83 and BLU-110 "dumb" bombs and 2,000-pound MK-84 and BLU 109 "dumb" bombs.[69] Based on the experience of Operation Desert Storm, it was meant to be a cost-effective solution to the requirement to enhance the accuracy of dumb bombs.[70]

The reform environment of the 1990s was multivaried. Its foundational concepts were CMI and price-based acquisition (PBA), which aimed to emulate many of the approaches and incentives of the commercial marketplace in defense acquisition. It included requirements reform, a key part of which was military specifications (MILSPECs) reform (which advocated the use of commercial and performance standards whenever possible) and required defense programs (such as JDAM and, later, SDB) to provide special justifications were MILSPECs used. A policy concept called *Cost as an Independent Variable* (CAIV) explicitly raised the priority of cost to the same level as system performance and schedule. This reorientation of cost as a lead priority differed from the traditional performance requirements-based military acquisition process, which had fewer inherent incentives motivating contractors to reduce cost.[71]

The overall acquisition reform movement of the 1990s drew heavily on the much-broader concept of CMI, which aimed to integrate commercial approaches, processes, technologies, and firms into defense acquisition.[72] Proponents of CMI, including the JDAM program SPO, were convinced that promoting the use of commercial parts and technologies and encouraging the participation of commercial firms would reduce technical risks at MS B, reduce costs, and raise quality, as well as enable PBA and CAIV approaches. Key elements of this approach as applied to the JDAM pilot program included the following:

- The JDAM SPO formulated clear, stable, and realistic system requirements.
- Competing contractors were required to conduct trade studies, commercial parts testing, prototyping, and extensive technology demonstrations prior to MS B.
- There was a heavy focus on incorporating the maximum amount of COTS parts, components, and technology.
- Competing contractors were granted configuration control to achieve performance outcomes according to their own design approaches.
- Contractors had to commit to an average unit production price (AUPP) and Production Price Commitment Curves (PPCC) prior to down-select at MS B.
- Competitive fixed-price production commitments were made during research and development.

---

[69] Lorell et al., 2000, p. 140.

[70] Although JDAM was a joint Air Force–Navy program, it experienced few of the organizational challenges of many joint programs in part because the Air Force so dominated the development effort. The program office was located at the Air Force Air Armaments Center at Eglin Air Force Base, Florida.

[71] Lorell et al., 2000, pp. 19–23.

[72] Lorell et al., 2000, pp. 1–3.

- There was little to no overlap across risk reduction, engineering, and manufacturing development (EMD), and production phases. Prototyping prior to MS B and during development helps ensure that a production-ready design was available at MS C.

Figure 2.4 shows the major events in the JDAM acquisition time line.

Figure 2.4. Time Line of Major JDAM Production Events

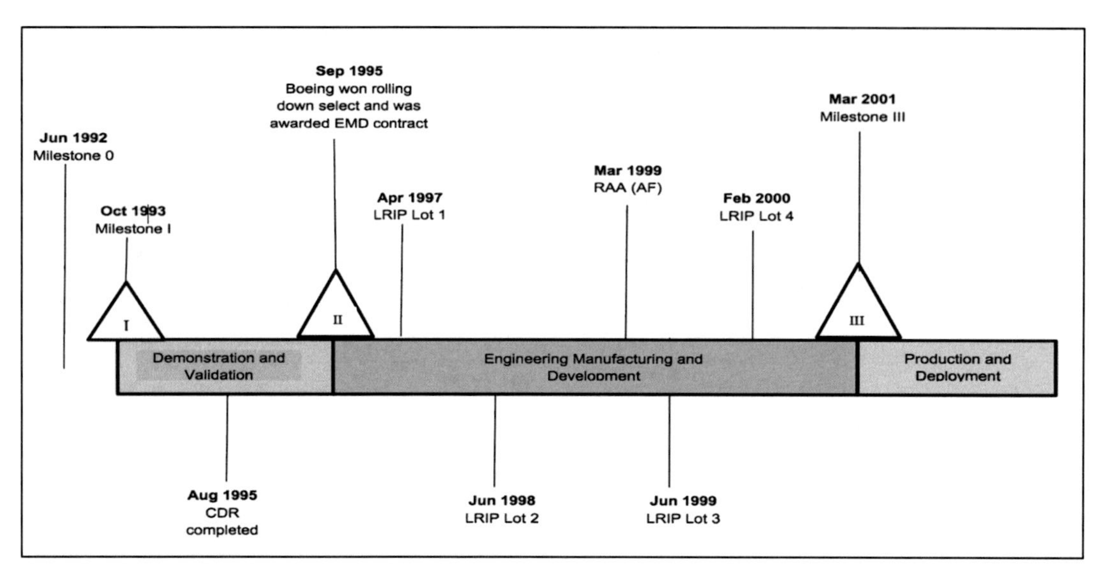

SOURCE: Based on information from JDAM SARs.
NOTES: RAA = Required Assets Available. When JDAM got underway, the milestones were labeled "Milestone I, II, and III," which is reflected in this figure. It should now be read as Milestone I = Milestone A; Milestone II = Milestone B; Milestone III = Milestone C.

JDAM Program Characteristics

Technology Maturity and Integration Complexity

Relieved of the requirement to use solely MILSPEC parts, competing contractors in the pre–MS B phase focused their efforts on integrating cost-effective, mature commercial, or government of-the-shelf (GOTS) parts into JDAM. Boeing and Lockheed Martin conducted affordability trade studies in addition to extensive testing of commercial parts. For example, according to one account, the use of commercial standard plastic encapsulated integrated circuits saved $535 per JDAM unit in BY dollars.[73]

Mature technologies were sought during the competitive advanced design (CAD) risk-reduction phase prior to MS B. In addition to trade studies and commercial parts testing, extensive prototyping and technology demonstrations were undertaken.[74] With JDAM, the plan

[73] Lorell et al., 2000, p. 155.

[74] Lorell et al., 2000, p. 155.

was to ensure technological maturity and identity problems early in the process to allow for a smooth transition to the EMD phase (now called the SDD phase). For example, tail-kit components and prototypes were developed and demonstrated in a research environment several years before the full-scale development program began.[75] Contractors completed pre–MS B prototypes and assessed and integrated various technologies, thus avoiding system integration problems down the road.[76]

With contractor configuration control granted during the CAD phase, Boeing and Lockheed settled on different mixes of commercial and military parts for their test prototypes in their attempt to find the best solution that maximized performance and minimized costs. For example, Boeing chose to integrate a military GPS receiver in its prototype, while Lockheed chose a commercial one. Out of the 13 key subcomponents integrated into Boeing's JDAM design, eight were commercial, and five were military. Lockheed incorporated ten components in its design: three commercial, four military, and three "hybrid" commercial/military components (see Table 2.2).

The EMD phase included two subphases. Phase I focused on further lowering technical risks and reducing unit costs.[77] It also included an 18-month "rolling down-select," during which the government provided competing contractors with detailed report cards every six months. These report cards included feedback on the strengths and weakness of each contractor's design proposal. The purpose was to enhance each contractor's design and thus promote even greater competitive pressure for the final down-select. Boeing won the Phase I rolling down-select in September 1995, with a final AUPP commitment from $14,000 to $15,000, compared with the government's AUPP requirement of $40,000 or less.[78] In a sense, Phase I was a competitive-technology demonstration and risk-reduction phase. Phase II entailed full-scale development, test, and demonstration of the weapon system by the winning contractor.

Requirements Realism and Stability

A key element of the new acquisition initiatives of the 1990s was the replacement of detailed technical requirements based on MILSPECs with performance requirements. Rather than providing the contractor with precise technical approaches, the services were encouraged to move toward using end-goal system performance requirements and provide the contractor with

---

[75] U.S. Government Accountability Office, *Defense Acquisitions: Strong Leadership Is Key to Planning and Executing Stable Weapon Programs*, GAO-10-522, May 2010b, p. 20.

[76] GAO, 2010b, p. 19.

[77] Lorell et al., 2000, p. 159.

[78] AUPP is the negotiated average price the government expects to pay for a JDAM weapon kit throughout an agreed-upon total quantity.

considerable flexibility to develop specific technical means of achieving the performance objectives.

Table 2.2. Commercial/Military Mix of JDAM Production Parts and Components

Item	Boeing	Lockheed Martin
Integration/assembly	Commercial	Military
Inertial measurement unit (IMU)	Military	Military
GPS receiver	Military	Commercial
Mission computer	Commercial	N/A
Circuit cards	Commercial	N/A
Connector	Commercial	N/A
Actuators	Commercial	Military
Power supply/distributor	Military	Commercial
Thermal	Military	Military
Container	Commercial	Military/Commercial
Fin	Commercial	Commercial
Tail	Military	Military/Commercial
Hardback/nose	Commercial	Military/Commercial

SOURCE: Lorell et al., 2000.

A cornerstone of JDAM program stability was the clearly articulated performance requirements, the most important of which were referred to as key performance parameters (KPPs). Extensive communication between JDAM program management, the user community, DoD, and the Air Force informed the trade-off analyses conducted during the initial requirements development process.[79] All stakeholders agreed on seven nontradable KPPs: weather capability, accuracy, in-flight retargeting capability, warhead compatibility, carrier operability, primary aircraft compatibility, and an AUPP of no more than $40,000, which was the most essential. All other performance attributes were tradable against cost and schedule. Contractors were not instructed on how to achieve the KPPs, thus allowing them to experiment with the most cost-effective method. Contractors were granted configuration control and freedom to experiment with commercial technologies and design approaches to meet or beat the $40,000 AUPP target.[80] The SPO wanted contractors to focus on mission performance and affordability rather than government-formulated detailed design and technical specifications.[81] This was expected to save time and contribute to a more innovative and cost-effective outcome.

---

[79] Lorell et al., 2000, p. 150.

[80] Lorell et al., 2000, p. 150.

[81] Lorell et al., 2000, p. 146.

This approach of focusing on key *capability* requirements rather than specific *technical* requirements combined with configuration control provided the contractors with considerable design and technical flexibility. It sought to demonstrate that cost and performance requirements could be achieved if strong communication and buy-in existed among all relevant parties. This was borne out as users and contractors adjusted their requirements and designs to complement each other and achieve the most cost-effective solution.[82] It also validated the departure from the more traditional approach, which relied heavily on MILSPEC requirements.[83]

Cost-Estimating Realism

In part because the departure from standard acquisition practices came after the program was underway, the SPO initially overestimated program costs. The SPO had an original APUC estimate of $68,000, but believed the weapon could be built for $40,000. Boeing's initial estimate was that JDAM could realistically be made for approximately $28,000.[84] However, competition between the contractors, continued collaboration within integrated product teams (IPTs) to integrate COTS and GOTS technology, and the feedback process during the rolling down-select worked favorably to drive down Boeing's and Lockheed's initial estimates. With a final AUPP commitment of $14,000 to $15,000, which was well below the government's original estimate, Boeing was awarded a firm fixed price (FFP) contract for LRIP lots 1 and 2 and a negotiated PPCC that covered lots 3 through 5 (a total of some 8,700 units).[85] A second PPCC for lots 6 through 11 was later established. Lot 6 came with a slight price reduction, with continuous price reductions extending to Lot 11. At the time, Boeing was the first defense industry company to commit to use a PPCC. Following a price-based commercial-like approach, Boeing calculated that offering price reductions for lots 6 through 11 would help its business strategy to win the rolling down-select. Thus these PPCCs were mutually beneficial for both SPO and contractor.

Acquisition Strategy and Program Structure

As an ACAT ID and Defense Acquisition Pilot Program (DAPP), JDAM's acquisition strategy has been described by analysts and government officials as "revolutionary," "commercial-like,"

---

[82] GAO, 2010b, p. 17.

[83] An example of MILSPECs that were retained despite the focus on commercial specifications is Mil-Std 1553 (inclusion of a high-speed bus through which the JDAM onboard processor communicated with the host aircraft) and Mil-Std 1760 (sets a standard for how JDAM's software interfaces with the host aircraft).

[84] No information is available on Lockheed Martin's internal cost estimates. However, Lockheed's best and final offer in the competition proposed a slightly higher APUC.

[85] Roger J. Witek, *Hard Skills, Soft Skills, Savviness and Discipline—Recommendations for Successful Acquisition: Case Studies of Selected Boeing Weapons Programs*, Montgomery, Ala.: Maxwell Air Force Base, Air University, AU/AFF/013/2008-05, May 2008, p. 35; and Lorell et al., 2000, pp. 60, 164–165.

and "price-based."[86] As mentioned, JDAM came to fruition within a new environment of reduced regulation.

After being declared a DAPP in 1994,[87] JDAM immediately experienced regulatory relief in the form of expedited waivers from various Federal Acquisition Regulations (FARs) and the Defense Federal Acquisition Regulation Supplements. Specifically, JDAM was granted 25 Truth in Negotiations Act (TINA)[88] and other waivers, and Contract Data Requirements Lists were reduced from 250 to 28. JDAM also enjoyed a reduction in MILSPECs compared with more traditional programs. The baseline pre-DAPP request for proposals (RFPs) included 87 MILSPECs, but this was reduced to a handful in the DAPP phase. The number of pages in the RFP also dropped from 986 to 285. The JDAM SPO did not require any specific commercial specifications or standards, nor were any MILSPECs or commercial standards embedded in the statement of work (SOW) because only a statement of objectives (SOOs) was required from the competing contractors. Switching from a SOW to SOO reduced the page count from 137 to two.

A carrot-and-stick approach (see Table 2.3) was adopted to incentivize AUPP and PPCC compliance. Many considered the "carrots" to be highly innovative in nature, like other aspects of JDAM's acquisition strategy. For example, if the contractor achieved all cost and performance goals,[89] it did not have to provide certified cost and pricing data per TINA to the government. The "sticks" were designed to protect the government from substandard contractor performance, especially related to cost and system performance. For example, such protective mechanisms would be enacted if the contractor submitted a price bid for a production contract lot that

---

[86] The Federal Acquisition Reform Act of 1995 established the statutory authority for OSD to designate DAPPs to test out new acquisition policy initiatives. The services submitted a list of candidate programs to OSD from which the USD(AT&L) selected the final list, which then had to be approved by Congress. DAPPs by law were relieved from the need to comply with all acquisition regulations and procedures, unless they were required by statute. However, each normal regulation or procedure that was deferred required a formal waiver approved by Congress. The USD(AT&L) established the final selection criteria for DAPPs. However, by statute, DAPPs had to be fully funded, based on an approved formal requirement, and deemed to be low risk. For more information, see U.S. Code of Federal Regulations, Title 32, Section 2.4, Designation of Participating Programs: National Defense, Washington, D.C.: U.S. Government Publishing Office, July 1, 2001. It is important to note that neither C-5 RERP nor WGS benefited from the special regulatory relief granted JDAM and SDB I. These two programs may not have performed as well if they had not been designated as DAPPs. However, designation as a DAPP did not necessary guarantee success. SBIRS High, one of the highest cost-growth programs examined in our companion document, had its single most important element, the Geosynchronous Earth Orbit (GEO) satellites, designated as a DAPP.

[87] Witek, 2008, pp. 30–31.

[88] TINA was enacted in 1962 (see Public Law 87-653, Truth in Negotiations Act, September 10, 1962). The law required the government to verify and vet all cost and pricing data associated with each government contract. It was seen as a burdensome accounting and reporting system. TINA waivers were granted in special circumstances and exempted the requirement for certified cost or pricing data, which allowed for a quicker acquisition process.

[89] In the 1990s, TINA data were considered by many acquisition reformers to be a burdensome and costly reporting requirement placed on contractors that inhibited commercial firms from competing for defense contracts and reduced contractor flexibility and incentives to innovate and take risks.

exceeded the PPCC for previously negotiated lots.[90] Table 2.3 provides a synopsis of the full range of carrots and sticks. These incentives helped encouraged the contractor to remain focused on cost reduction and were a core component of JDAM's acquisition strategy.

Table 2.3. JDAM PPCC Carrot-and-Stick Approach

Carrot	Stick
Contractor remains the sole production source for an agreed number of lots if PPCC and AUPP commitment is achieved.	If not, contractor must fully qualify, at his or her own expense and within 12 months, a new contractor as a second source for production.
Contractor retains full configuration control as long as changes do not reduce performance or impact safety of flight.	Government can reestablish control over configuration if product performance is substandard.
Contractor can retain savings or any additional profit if it is able to insert new technologies and reduce production costs if KPPs are achieved.	If not, contractor is required to prepare and provide a fully compliant MILSPEC data package to government.
Contractor will not be required to submit certified TINA cost or technical data to the government if performance, reliability, and delivery schedules are met.	If not, contractor must submit fully compliant certified cost and pricing data in accordance with TINA and other regulations.
No in-plant government oversight or inspection of the contractor or subcontractors.	Government may impose in-plant oversight and testing.
Contractor receives an incentive fee if the accuracy and reliability of production units exceed the specification.	Government eliminates the incentive fee option.

SOURCES: Lorell et al., 2000; and Witek, 2008.

The competing JDAM contractors embraced PBA. With traditional cost-plus development programs, system engineers strive to design the best product with relatively less concern for cost. However, in JDAM and other acquisition pilot programs of the 1990s, contractors had to design to specific affordability goals at the outset to meet or beat specific government cost goals. This resulted in a highly iterative development process, where system integration and producibility were continuously assessed. Sole-source production was awarded to the more-compliant contractor. AUPPs, PPCCs,[91] and price-quantity production matrices were negotiated well before MS B. PPCC and FFP contracts encouraged competitive but realistic pricing between Boeing and Lockheed Martin, and the collaborative nature of EMD and the rolling down-select[92] process

[90] Lorell et al., 2000, pp. 165–167.

[91] With PPCCs, contractors commit to a specific pricing curve over a specific number of lots at the beginning of full-scale development. The purpose is to guarantee production prices and system performance. With PPCC, contractors customarily agree to lower unit prices for high production/volume orders, allowing the government to save money. PPCCs are intended to be good-faith best estimates and are not contractually binding.

[92] "Rolling down-select" was a key component of the acquisition reform environment. The concept implied that constant interaction would occur between contractors and the government to help contractors identify weaknesses in their proposals, and to present the best possible offer for the government prior to down-selection. IPTs helped facilitate such collaboration.

promoted both transparency and realism while intensifying contractor competition.[93] Recognizing that honesty and cost realism were key, contractors avoided underbidding and overpromising to earn contract incentives. They also engaged in careful supplier management to ensure subcontractors could meet AUPP requirements. Boeing worked diligently to persuade its suppliers to accept price reductions for lots 6 through 11, especially because these lots saw a substantial increase in production numbers.

IPTs were integral to JDAM's acquisition process.[94] The IPTs analyzed each JDAM component to find cost drivers. Contractors gave their suppliers and subsuppliers price targets with continual updates, and subsuppliers were allowed to participate in performance-specification implementation.[95] This process of mutual analysis by subsuppliers increased transparency within the IPTs and helped drive affordability.

Milestone B/C

JDAM successfully avoided excessive overlap of the development and production phases by adhering to a sequentially phased schedule and competitive testing before MS B. JDAM did have several relatively minor technical and early flight-test problems, including relatively minor problems with the radio frequency components, GPS, system vibration, and friction brake.[96] However, delaying Full FRP for approximately a year and restructuring developmental and operational testing mitigated many of these problems, thereby avoiding launching LRIP with immature or unstable designs.[97]

---

[93] Other programs attempted implementing various aspects of this approach, with less success. More is said on this in the concluding chapter.

[94] There were three main all-government IPTS. Each contractor was assigned its own IPT, which had no access or insight into the other contractor's IPT. These IPTs worked very closely with the contractor to develop the best-possible design proposal for the government and were specifically instructed to help "their" contractor to win the competition. A third Overarching IPT had exclusive access to information and results from each of the two IPTs assigned to the contractors and could thus oversee and promote the entire process. The goal of the IPTs was to achieve the situation where each contractor would provide the best possible proposal at the end of the rolling down-select, thus providing the government with the best possible options to chose from.

[95] Witek, 2008, p. 33.

[96] Lorell and Graser, 2001, pp. 72–73.

[97] Lorell and Graser, 2001, p. 76. It is important to note that this type of careful sequencing of acquisition phases combined with acceptance of schedule slippage to ensure that prior phases are complete is only possible if adequate time is available to implement it. If there is an urgent need or a sudden change in the schedule requirement for important national security reasons, programs may not have the luxury of implementing such a strategy.

JDAM Summary of Findings

JDAM was an important 1990s acquisition reform Pilot Program that came to fruition in an acquisition environment stressing a novel acquisition strategy and reduced regulatory oversight. Various components of this new environment contributed to JDAM's ability to stay on schedule and achieve negative cost growth:

- A clear and stable requirement for an expeditiously developed and affordable guided bomb tail kit assembly providing increased accuracy, compared with previously used "dumb" bombs in Operation Desert Storm.
- Competitive prototyping combined with a rolling down-select that provided feedback to competing contractors prior to MS B.
- An acquisition strategy that stressed integration of mature, commercial technologies in a commercial-like CMI environment that was focused on PBA.
- Prepriced production lots prior to MS B that used price-quantity matrices based on contractor commitment to AUPP and a PPCC.
- Use of performance requirements, contractor configuration control, and carrots and sticks to adhere to the AUPP commitment and PPCC.

Given the role of JDAM as a pilot program for acquisition reform and many new and innovative acquisition approaches, and the fact that it experienced negative PAUC growth, it is not surprising that JDAM experienced none of the five common characteristics that we found in programs with extreme cost growth. Instead, it had the opposite characteristics (shown in Table 2.4). JDAM had a rigorous preparatory period making sure it was ready for MS B and full-scale development. This period included intensive competition between two prime contractors who engaged in extensive prototyping to validate their designs and reduce technical uncertainties and risk. In addition, JDAM is a relatively less-complex system, with fewer than ten major subcomponents, many of which are based on well-known existing technologies and components.[98]

[98] Four components—the IMU, GPS receiver, and control actuators make up 85 percent of the production cost of the JDAM (see Lorell and Graser, 2001, p. 48).

Table 2.4. JDAM Had None of the Common Characteristics of the Six MDAPs with Extreme Cost Growth

	JDAM
Premature MS B	
Immature technology; integration complexity	
Unclear, unstable, or unrealistic requirements	
Unrealistic cost estimates	
Acquisition policy and program structure	
Acquisition strategy and program structure not tailored for level of risk	
MS B/C	
PAUC growth (%)	–12

JDAM also had clear, stable, and realistic requirements, as well as realistic but highly aggressive cost estimates. Program officials experimented with a much more capability-based (rather than specification-based) requirements process, which permitted the contractors to experiments with differing technical approaches. A small number of KPPs were specified, including a KPP for AUPP, rather than detailed technical specifications. Given the relative simplicity of the system and the well-known nature of the component technologies and subsystems, accurate cost estimate was certainly feasible. Interestingly, the program baseline estimates, as well as the unit-cost KPP, were very aggressive, as incentives to the contractors to lower costs.

Because JDAM was an official DAPP testing out many new acquisition approaches, it is not surprising that the acquisition strategy and programs structure were carefully tailored for this specific effort and the relatively low level of technological risk inherent in the program. This included waivers for many normal acquisition regulatory and reporting functions. Interestingly, although the program was relatively low risk, planners completely rejected the common approach in the 1990s of adopting a simultaneous MS B/C approval point to save time and money. This was likely a sound decision, given that nearly all programs that adopted a combined MS B/C approach eventually experienced difficulties. Indeed, JDAM's initial schedule envisioned more than five years of development and testing between MS B and MS C.[99]

---

[99] DoD, *Selected Acquisition Reports, Joint Direct Attack Munition (JDAM)*, Washington, D.C., December 2014c.

Small Diameter Bomb I

Figure 2.5. GBU-39 SDB I

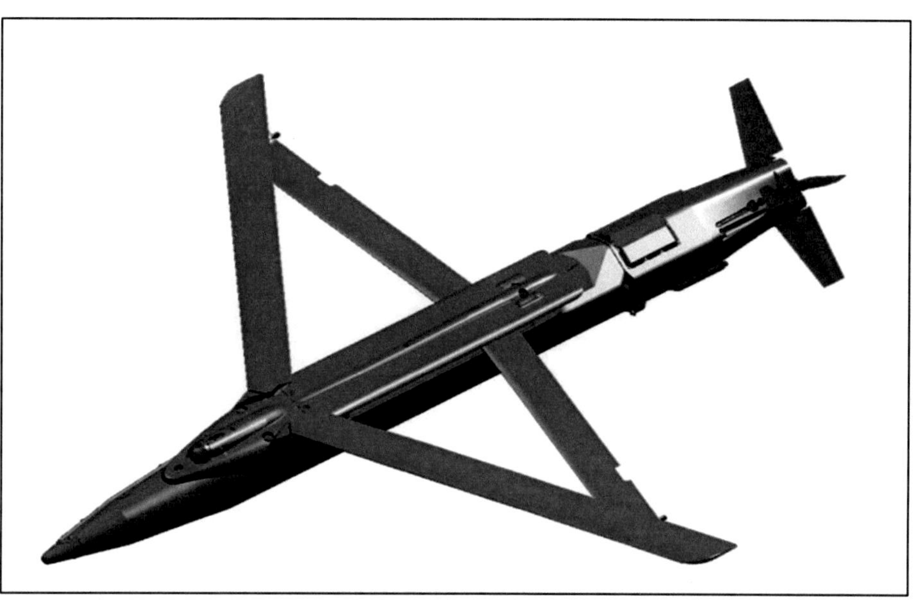

SOURCE: Courtesy of U.S. Air Force.

Summary Overview

Boeing's SDB I is an Air Force ACAT ID program designed to provide warfighters an affordable, all-weather precision-munitions capability against stationary targets. A key aspect of the requirement is to reduce the size of the weapon to provide a much more numerous weapon load-out on attack aircraft. Fielded in 2006, SDB I is a 250-pound "smart" bomb that can travel more than 60 nautical miles while using GPS technology to ensure accuracy, similar to JDAM. Its Advanced Anti-Jam GPS-aided INS guides the weapon to the coordinates of a fixed target.[100] Compared with JDAM, which is strap-on tail kit affixed to a pre-existing dumb bomb, SDB is a more complex, all-in-one singular miniaturized smart munition that includes retractable wings. The first increment of SDB (originally called SDB Phase I)[101] was designated for carriage on the F-15E, with a long-term goal of F-22 and F-35 compatibility.[102]

SDB and JDAM program management shared the same strategic objective—to quickly field an affordable, operationally effective, accurate, and reliable weapon. Both programs were

---

[100] Boeing, "Backgrounder: Small Diameter Bomb Increment I (SDB I)," January 2012.

[101] The SDB program was originally constructed to consist of two parts (or phases): SDB I and SDB II. SDB I, the first phase, would seek stationary/fixed targets, while SDB II would seek moving/mobile targets. SDB was eventually split into separate programs, SDB I and SDB II. SDB 1 production ended in 2013 and SDB II began soon after, with MS C potentially occurring in May 2015 (according to the 2014 SDB II SAR).

[102] DoD, *Selected Acquisition Reports: Small Diameter Bomb I (SDB-I)*, Washington, D.C., December 2005c, p. 3.

acquisition pilot programs expected to navigate a newly streamlined acquisition process by focusing on such elements as CMI and reduced schedule length. The SDB SPO reportedly placed a greater emphasis on schedule compared with JDAM officials who stressed the importance of low cost. Despite this difference, both evolved into successful low cost-growth programs for similar reasons, many of which stem from the 1990s acquisition reform environment. As with JDAM, regulatory oversight was reduced and comprehensive contractor incentives were implemented to achieve performance and cost goals.

From the onset, SDBs, like JDAM, adopted such unusual measures as using a commercial-like approach to acquisition, integrating mature commercial parts and components in prototype designs, aggressively competing prototypes from multiple contractors prior to down-select for full-scale development, testing often and early, and striving for stable requirements and early production pricing commitments from contractors. All of the above were intended to occur prior to source selection and MS B. Air Combat Command, OSD, the competing contractors, and other stakeholders supported an acquisition strategy that complemented these objectives with contract and production incentives. In addition to SDB's wide use of COTS and its integration and testing successes, contractors also agreed to AUPPs prior to final down-select, which helped ensure the production of a low-cost, high-quality reliable and maintainable system. The SPO also required competing contractors to provide PPCCs and long-term pricing agreements with a single fixed prenegotiated price within a range of quantities in each lot, which helped keep costs predictable and low during production.

AUPP and PPCC were new concepts in the 1990s that were first used on JDAM and used carrot-and-stick approaches to guarantee affordability. Unlike JDAM, the competing SDB I contractors agreed well before MS B that lot prices would remain constant no matter what quantity was ordered, as long as the quantity remained within prenegotiated minimum or maximum numbers.[103] Figure 2.6 indicates the major events in the SDB I acquisition time line.

---

[103] Lorell et al., 2000, p. 163.

Figure 2.6. Time Line of Major SDB I Production Events

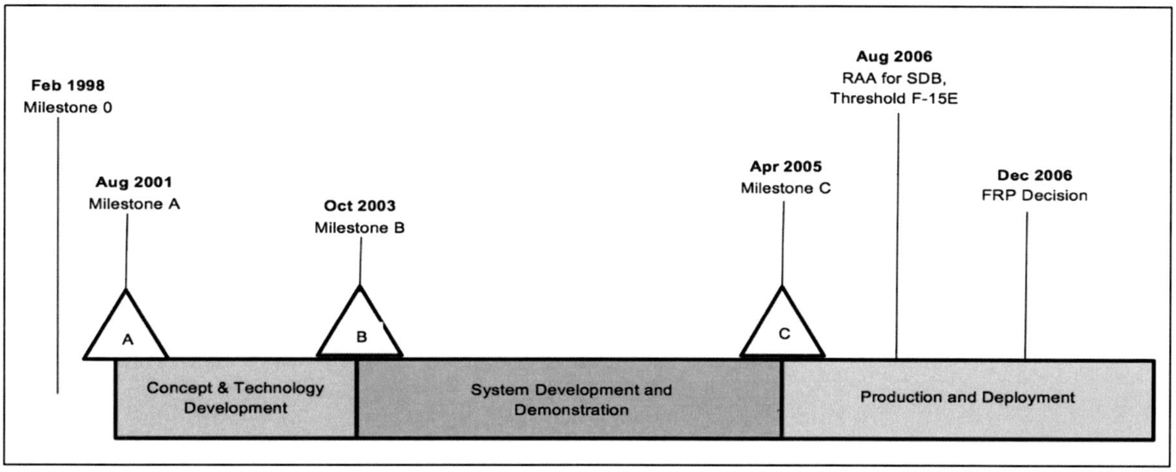

Feb 1998
Milestone 0
Aug 2001
Milestone A
Oct 2003
Milestone B
Apr 2005
Milestone C
Aug 2006
RAA for SDB,
Threshold F-15E
Dec 2006
FRP Decision
A
B
C
Concept & Technology Development
System Development and Demonstration
Production and Deployment

SOURCE: RAND graphic based on data from SDB I SARs.

SDB I Program Characteristics

Technology Maturity and Integration Complexity

Despite being somewhat more complex than JDAM, SDB I can still be characterized as a technologically modest development program. From the outset, competing contractors aggressively sought mature COTS and GOTS technology for integration into prototypes. COTS technology, in particular, was to be tested early and often to lower risk and ensure compatibility before integration.[104] For example, the SDB guidance system, warhead, and communications link technologies were developed prior to program start. The Air Armament Center supported this early and aggressive testing schedule, and competing contractors were cognizant of heightened expectations during the two-year competitive CAD phase.

Thus, there was a strong commitment to obtaining system technology maturity prior to MS B. According to a 2010 GAO report, contractors were given credit toward meeting the down-select criteria only for demonstrated (as opposed to aspirational) performance of their prototypes. This meant that a results-based approach to systems engineering was adopted for production-representative hardware, which program management saw as essential. Contractors only received credit if they demonstrated performance greater than the threshold requirement. They did not receive credit, however, for promising greater performance in future spirals.[105] The SPO stressed that once the program entered the system development phase, the winning contractor should only have to focus on building more units, readying the factory floor for LRIP, and completing testing. In other words, by MS B, the program office expected the contractor to have developed

[104] Lorell et al., 2000, p. 163.

[105] Witek, 2008, p. 61.

"production representative hardware that met stated requirements," and that would be fully ready at MS C with stable reliable design for full-scale production.[106]

Other examples of pre-MS B technology readiness measures that both contractors carried out include finalized design reviews, early live-fire tests, more than 80 percent of production hardware flight tested, and six flight tests (Boeing).[107]

Requirements Realism and Stability

SDB was designated as the Air Force's highest-priority program at Edwards Air Force Base Air Armaments Center, which also housed SDB's program office.[108] Top-down guidance emphasizing SDB's high priority established a pattern of strong communication that permeated the program—resulting in such stakeholders as warfighters and combatant commanders "falling in line" behind the SDB's core objectives. Early in the planning for SDB, the Chief of Staff of the Air Force (CSAF) produced a clearly written "Commander's Intent" stating that his priority was to have the weapon system available for use by fourth quarter (4Q) FY 2006, according to the GAO.[109] Program officials claimed that having this priority from the top was invaluable in spurring the SPO to develop an effective business plan for acquiring the system on time.[110] As with JDAM, system requirements development and acquisition planning could be streamlined because of the DAPP status of the program. The SPO ensured that stakeholders and the competing contractors stayed focused on the most essential system requirements and avoided adding additional ones down the road. Avoiding requirements "creep" was viewed as a fundamental necessity if SDB was to adhere to schedule and cost objectives.

The SPO clearly communicated to contractors which SDB program requirements were fixed versus tradable. The program specified a maximum of two fixed or mandatory requirements or KPPs: weapon load-out[111] and GPS interoperability. Other requirements (e.g., accuracy, effectiveness, reliability) had thresholds and objectives that were deemed tradable.[112] Program management was motivated to prioritize among the competing demands of cost, schedule, and technical performance and to find an "80% solution that could help the warfighter." To do this, the SPO focused relevant stakeholders on meeting their commitments and keeping to schedule;

---

[106] Witek, 2008, p. 61.

[107] DoD, 2003c, p. 4.

[108] DoD, 2003c, p. 11.

[109] DoD, 2003c, p. 11.

[110] DoD, 2003c, p. 11.

[111] *Weapon load-out* refers to the required carriage, configuration, and quantities of ordnance for a specific type of aircraft.

[112] A *threshold requirement* is a minimal acceptable value for a given parameter or specification, whereas an *objective requirement* is the more desirable or aspirational value (see Witek, 2008, pp. 55–56).

warfighters were encouraged to postpone submission of new requirements to avoid requirements creep.[113] For example, the Army component of the Joint Chiefs of Staff argued that SDB should be EMP-hardened as a protection against nuclear attack; however, the SPO, with support from OSD, was able to stave off this additional requirement.[114] All along, SDB program management and OSD remained focused on what was needed to actualize its streamlined acquisition approach: Requirements were monitored to discern what was critical versus negotiable.

Cost-Estimate Realism

From the beginning, the Air Force stressed the importance of affordability for SDB I. The iterative trade-off approach that was applied to the requirements process was also applied to the cost-requirement process. The trade-off analysis for requirements helped increase the feasibility of a $30,000 AUPP threshold for competing contractors. Contractors refrained from unrealistically underbidding their work and overpromising on SDB performance. They understood that developing a low but realistic AUPPC would enhance their credibility during source selection.

Thus, SDB paralleled its predecessor JDAM in a number of ways. During Phase 2 of EMD for JDAM, sole-source selection primarily came down to a question of the lowest realistic production price commitment (once KPPs were met).[115] A series of contractual carrots and sticks designed to enforce adherence to the PPCC after down-select were used on both programs. With SDB, this carrot-and-stick approach—which the SPO referred to as a "stability incentive"—was used to guarantee the AUPP and PPCCs for both LRIP and follow-on production lots. However, SDB's carrot-and-stick approach departed somewhat from JDAM's in that it sought to incentivize schedule and design stability just as much as cost. The SPO expected contractors to design in high reliability, which was made possible by a focus on stable requirements and design stability from the start. The carrot was $5 million for each production lot, which retained design stability while still meeting all performance and reliability requirements. The stick was that this was an all-or-nothing incentive. Competing contractors understood that adherence to the stability incentive provided financial remuneration, which helped ensure compliance. The results were favorable. Boeing was incentivized to meet the $30,000 AUPP and ended up reducing development costs by almost 5 percent and unit costs by more than 14 percent—making the system available one month earlier than planned.[116]

---

[113] Witek, 2008, pp. 55–56.

[114] GAO, 2010b. p.11

[115] Lorell et al., 2000, p. 163.

[116] GAO, 2010b, p. 11. Lockheed was out of the competition by this time. We do not have information on Lockheed's final and best offer regarding development cost and AUPP.

Success with cost realism can also be attributed to SDB's results-based approach, which forced contractors to demonstrate real rather than aspirational system performance below a specific cost threshold. To enter SDD, detailed production drawings had to be at least 85 percent complete and the configuration control board had to be established and functioning. The drawings included carefully formulated criteria that made it difficult for contractors to underbid. And to ensure cost realism, the SPO required that contractors' estimates fall within 15 percent of the SPO's estimates for SDD.

These strategies allowed the SPO to avoid cost overruns during SDD or price hikes during production. As mentioned above, PPCCs and long-term pricing agreements during both LRIP and FRP helped stabilize prices. Prior to down-select, competing contractors were required to offer fixed prices for lots 1 and 2 and average unit price commitments for lots 3 through 7.[117] When Boeing was selected as the prime contractor, it was under the proviso that FFP contracts and an AUPP/PPCC approach would be the norm. Accordingly, Boeing agreed to a prenegotiated fixed price for each lot. This meant that, during a surge situation where the government contracts for a certain number of systems, Boeing would be precluded from raising the unit cost above the agreed price for that particular lot as long as the number remained below a prenegotiated maximum.[118] SDB program management explained that, when it did not, the prime contractor would be responsible for all costs associated with the "ramp-up" process (e.g., extra labor, capital investment, increased subcontractor output). This contrasts with JDAM, which had a fluctuating price-quantity matrix, where the government would pay higher per-unit prices for lower quantity production lots.[119] SDB, on the other hand, lacked a price quantity matrix and was beholden to a single AUPP and PPCC commitment, irrespective of unit quantity (within prenegotiated maximum and minimum quantity bounds).

Acquisition Strategy and Program Structure

SDB's acquisition strategy had four main cornerstones: incrementally phasing in low-risk technology, testing early and often prior to MS B, committing to key performance parameters, and using AUPPs and PPCCs to keep costs low. Together, these four components made for an

---

[117] Witek, 2008, p. 61.

[118] Unlike with JDAM, there was a single prenegotiated fixed price for each lot. However, there was also a maximum number and a minimum number limitation. If the government-ordered numbers above the prenegotiated maximum or below the prenegotiated minimum, Boeing could demand new negotiations for a new unit price.

[119] Interview with SDB program management office, November 12, 2010. Unlike SDB I, JDAM production contracts each have a large matrix of different quantities and different prices for each lot. In contrast, SDB I had one negotiated price for a large range of different quantities. However, there were limits at each end of the spectrum; that is, a minimum number and a maximum number, after which the contractor had a right to seek renegotiation of the contract. Interestingly, however, when JDAM experienced a major unanticipated production surge following the September 11 attacks, which greatly exceeded the maximum prenegotiated contract lot number, Boeing ultimately agreed to keep the price stable and not renegotiate a higher price even though they had the contractual right to do so. This was due to pressure from Congress, and according to Boeing, an attempt to regain good will with the customer.

effective acquisition strategy that had characteristics of a commercial-like business approach. This streamlined approach to acquisition also facilitated the CSAF's RAA objective of 4Q FY 2006.[120] To meet the RAA target, CSAF wanted to do as much component and total system testing upfront as possible.[121] Air Force officials touted the positive impact of the CSAF's establishing a clear priority among the competing demands of cost, schedule, and technical performance.[122]

The streamlined approach required that technology be mature while simultaneous tests were conducted to ensure compatibility. On account of this, integrating mature COTS and GOTS technology was an early objective after seeing how well it worked with JDAM. This entire strategy was enveloped within a highly communicative and interactive environment. To underscore this point, as with JDAM, numerous IPTs were established. Seen as a means of promoting innovation and timeliness, SDB's IPTs were integral to pre–MS B successes. Many of these IPTs were described as "hand-picked-A-teams" that routinely touted "go-fast-plans," which were designed to balance discipline and swiftness.[123] The participants understood the streamlined acquisition environment in which they worked. The IPTs' combined "experience enabled them to independently drive multiple fronts while staying connected to the program-level strategies."[124]OSD, the Air Force, the contractors, suppliers, subsuppliers, and warfighters— in a manner similar to JDAM—worked collaboratively and communicated their capabilities and limitations. Air Force officials credited these IPTs with fostering communication among SDB supply-chain components and Pentagon decisionmakers alike,[125] keeping all parties focused on the RAA 4Q FY 2006 objective.[126] SDB also only had a single program manager who served from before MS B through the low-rate production decision, providing leadership continuity for the entire system development phase.[127]

Lastly, the SPO and competing contractors agreed to an AUPP prior to down-select and to FFP contracts with PPCCs after source selection. At the time, PPCCs were considered a relatively new approach that complemented PBA. Using such long-term pricing agreements helped the SPO keep costs low and stable by requiring Boeing and its suppliers to shoulder any

---

[120] RAA is government driven and differs from initial operating capability, which is entirely dependent on contractor actions and independent of the government. RAA's can be contractually binding and come with carrots and sticks to incentivize the competing contractors accordingly.

[121] Witek, 2008, p. 62.

[122] Witek, 2008, pp. 55–56.

[123] Witek, 2008, pp. 56, 68, 94–99.

[124] Witek, 2008, pp. 55–56.

[125] Witek, 2008, pp. 94–99.

[126] Witek, 2008, pp. 56–57.

[127] GAO, 2010b, pp. 15–20.

additional costs that accrued from production surges, as long as the government stayed within the maximum and minimum numbers.

Milestone B/C

SDB's incremental approach to acquisition—where mature technology was tested early, often before full-scale development and prior to production—allowed for the identification of problems and the avoidance of concurrency.[128] The CSAF prioritized schedule and mandated an ambitious 4Q FY 2006 RAA, but not at the expense of excessive overlap of the development and production phases and guaranteeing a production-ready stable design at MS C.[129]

SDB I Summary of Findings

SDB I was designed to meet a pressing Air Force need for an accurate, low-cost, low–collateral damage weapon whose small size permits much-higher load-out on standard attack aircraft. It is considered a stable program that achieved negative cost growth in part because of its commercial-like acquisition strategies. Other factors contributing to SDB's low cost growth were:

- clear, realistic, well-defined, and stable requirements
- clear and consistent top-down guidance from senior Air Force and OSD leadership, which prioritized timeliness and affordability
- a trade-off analysis approach for system requirements, which mandated only two essential KPPs: weapon load-out and GPS interoperability, with all other requirements considered tradable based on cost-benefit analysis
- a competitive down-select process that required AUPP and PPCC commitments by competing contractors, who were thus motivated not to overpromise on SDB performance
- use of performance requirements, contractor configuration control, and carrots and sticks to adhere to the AUPP and PPCC commitments
- prepriced production lots with one lot price independent of lot quantity within a maximum and minimum quantity range.

Table 2.5 shows that SDB I, such as JDAM, also had none of the key characteristics common all the six MDAPS with extreme cost growth. This was true of only JDAM and SDB I, which are also the two low cost-growth MDAPs with negative cost growth. SDB I is the best performing of the four MDAPS with low cost growth, with PAUC growth of –16 percent, as shown in Table 2.5.

---

[128] GAO, 2010b, p. 28.

[129] Witek, 2008, pp. 56–57.

The SDB I acquisition strategy and approach were closely patterned on the JDAM example. In many respects, SDB I is similar to JDAM in that it is a GPS-guided winged unitary smart

Table 2.5. SDB I, with the Lowest PAUC Growth, Had None of the Key Characteristics of the Six MDAPs with Extreme Cost Growth

	SDB I
Premature MS B	
Immature technology; integration complexity	
Unclear, unstable, or unrealistic requirements	
Unrealistic cost estimates	
Acquisition policy and program structure	
Acquisition strategy and program structure not tailored for level of risk	
MS B/C	
PAUC growth (%)	–16

stand-off munition designed for use against high-value fixed target. JDAM has the same mission and uses the same basic technologies, but instead of a unitary winged munition, it serves as a strap-on kit for existing, much-larger dumb bombs. Given the fact that SDB I closely followed the same path pioneered by JDAM, it is not surprising that, like JDAM, SDB I also evince any of the characteristics common to the six MDAPs with extreme cost growth.

Wideband Global SATCOM (WGS) System

Figure 2.7. WGS System

SOURCE: Courtesy of U.S. Air Force Space Command.

Summary Overview

WGS, whose original name was Wideband Gapfiller Satellite, was initially intended to be a relatively low-cost, commercially derived interim "gapfiller" communications satellite system to augment DoD's Interim Wideband System, which included the Defense Satellite Communications System (DCSC) III and the Global Broadcast Service (GBS) Phase II. Originally, it was a jointly funded Air Force–Army program. Its central goal was to achieve an interim upgrade of the capabilities of the aging DCSC III and GBS systems pending the development of the AEHF satellite and the next-generation Objective X/Ka Wideband System or Advanced Wideband System, which was later called the Transformational Communications System, then the Transformational Satellite System (TSAT).[130]

In short, WGS was originally intended to be a short-term, temporary, low-cost supplement for an aging defense space communications system, which would later be permanently upgraded with much more capable new-generation communications satellites. The intended low-cost and low-risk commercial basis of the program is illustrated by the decision of DoD's Defense Acquisition Board in November 2000 to proceed directly into a combined development and

[130] See DoD, *Selected Acquisition Reports: Wideband Global SATCOM (WGS)*, Washington, D.C., December 2001a; and "Wideband Gapfiller System," GlobalSecurity.org, undated.

production phase (MS II/Production or B/C), with first launch scheduled for fewer than four years, in the second quarter of FY 2004. This is also confirmed by the contracting strategy; Boeing was awarded an FFP contract for six satellite options in January 2001 at MS B/C, although only three SVs or satellite vehicles were planned and funded at program inception, with development estimated at $175 million and total procurement of three SVs at $678 million. More telling, this contract was written under FAR Part 12 commercial item acquisition regulations. FAR Part 12 waives many of the normal military acquisition contractor reporting and oversight requirements to facilitate the acquisition of purely commercial items.

Studies conducted since the late 1990s by the government and Boeing have suggested that WGS could incorporate existing commercial communications satellite technologies to provide a low-cost temporary gapfiller for the military. For example, WGS was supposed to be compatible with existing ground terminals and stations, and would incorporate active phased array X-band antenna technology, digital signal processing technology, software algorithms, and commercial parts, processes, and procedures developed by Boeing for commercial communications satellites. The SV was to be based on the Boeing commercial 702 satellite bus.[131]

WGS ran into some difficulties when it evolved from a temporary gapfiller to a permanent solution for upgrading DoD's communications satellite systems. This transition resulted from a variety of causes: changes in budgetary priorities; issues involving the AEHF satellite program; problems with launch services scheduling; and because the ultimate next-generation satellite program, the TSAT, was eventually canceled because of growing costs and technical problems.

WGS program structure and history are more complex than the other three low cost-growth programs. To clarify our approach to the analysis of this case, a longer introductory section is required compared with the other three cases. This necessitates a more-detailed overview of the entire program and a brief explanation of the RAND methodology for assessing cost growth on WGS.

Detailed Overview of the Entire WGS Program

The WGS program is different from many other MDAPs in that its purpose and scale changed significantly over the course of the program. As noted above, the program was originally intended to be a temporary stop-gap measure to upgrade the aging military space

---

[131] DoD, December 2001a; "Wideband Gapfiller System," undated. A satellite or space bus is the basic SV or structural platform that carries and supports the scientific- or mission-specific payload module. Typical space buses provide electric power, propulsion, communications equipment, attitude control, guidance, navigation and control equipment, and so forth. The payload module is carried on the bus. Since all satellites require these support capabilities, some types of satellites can use the same or similar bus to support different types of customized mission payloads. Use of a common, existing, or commercial bus can reduce nonrecurring costs of developing a specialized bus for every type of mission payload (Committee on Earth Studies, Space Studies Board, Commission on Physical Sciences, Mathematics, and Applications, National Research Council, *The Role of Small Satellites in NASA and NOAA Earth Observation Programs*, Washington, D.C.: National Academy Press, 2000).

communications system with three advanced but low-cost commercially derived SVs pending the development and launch of AEHF satellites and the definitive next-generation military communications satellite, the TSAT. While the initial FFP contract to Boeing included options for six satellites, the baseline plan and funding assumed only three satellites, one planned for launch in 2004 and two in 2005.

By the end of 2002, however, a variety of internal and external factors began affecting the program. These factors simultaneously delayed the planned March 2004 launch date for WGS SV 1 and increased the need for more WGS SVs. The first delay was caused by growing manufacturing problems on WGS, which required more time and greater budgetary resources to mitigate. Budget reductions also caused delays. In May 2002, DoD requested a reprogramming of more than $88 million from WGS to SBIRS to help that troubled program. This loss of funding caused a six- to seven-month delay in the scheduled launch of the first WGS SV.[132]

More importantly, WGS, despite its commercial lineage, began experiencing several developmental and manufacturing challenges. Initially, software in the control element experienced some problems, which resulted in delays.[133] Then, in early 2003, Boeing began encountering developmental and manufacturing problems with a key subsystem, the phased array antenna. This initially led to an additional five-month delay, resulting in a schedule slip of nearly one year to February 2005 for the first launch.[134]

These WGS delays, combined with developmental, funding, and launch uncertainties for both AEHF and TSAT, plus the need to ensure the viability of the military satellite communications system, led DoD to approve two of the additional unfunded WGS production options for a total of five SVs. Because of budgetary, funding, and other constraints, the fourth and fifth WGS SV options could not actually be exercised within the contractually required 24-month period ending in November 2002, following the exercise of the third WGS option. This meant that the original FFP per satellite negotiated at MS B/C was no longer binding, and Boeing could open negotiations for a new contract and new price.

DoD and the Air Force recognized that this price would likely be considerably higher for several reasons. First, as noted above, Boeing was experiencing developmental issues and manufacturing delays that were raising its costs. Under the original fixed-price contract, Boeing was likely losing money. Second, DoD now wanted to add additional wideband capabilities to the remaining WGS SVs to accommodate rapidly growing RPV data communication requirements for airborne intelligence, surveillance, and reconnaissance (ISR). The initial goal was to at least double the available bandwidth of two Ka-band antennae to support RPV

---

[132] Amy Butler, "Wideband Gapfiller Launch Raided to Pay for FY02 SBIRS High Bailout," *Inside Missile Defense*, August 21, 2002.

[133] DoD, *Selected Acquisition Reports: Wideband Global SATCOM (WGS)*, Washington, D.C., December 2002a.

[134] DoD, *Selected Acquisition Reports: Wideband Global SATCOM (WGS)*, Washington, D.C., December 2003a; and Amy Butler, "Wideband Gapfiller Satellite to Launch Nearly a Year Late," *Inside Defense*, August 1, 2003.

communications, also called a "radio frequency bypass capability," which would more than triple bandwidth for RPV data requirements. Third, an anticipated increase in the commercial satellite market failed to materialize, threatening vendor survival and raising the costs of parts. Fourth, the original MS B cost estimate was too optimistic from the beginning and underestimated the difficulty and complexity of developing and producing the commercially derived WGS. [135]

DoD was unable to award the contract for the additional WGS SVs and exercise the option for WGS SV 4 until November 2006. In the time since the initial decision to expand the buy and renegotiate the contract, Boeing had encountered further manufacturing and technical challenges, which, in addition to other factors, continued to drive up the price. An additional three-month delay in 2005 was caused by changes in schedule and launch vehicle availability because of a higher-priority National Reconnaissance Office payload.[136] This was followed by growing manufacturing technology problems at Boeing. For example, Boeing discovered that fasteners on WGS SV 1 had been improperly installed. As a result, considerable time was spent replacing the fasteners on SV 1 and inspecting SV 2 and SV 3 for the same problem. Other manufacturing and technology issues arose with a variety of parts on the phased array antenna. Originally expected to cause an 18-month delay, these issues and the earlier problems ultimately led to a three-and-a-half year slip in the launch schedule for WGS SV 1, which was finally launched successfully in October 2007.

While Boeing had to absorb all the cost growth on SVs 1 through 3 because of the FFP contract under FAR Part 12 rules for commercial procurement, the contract for the next group of satellites (now called Block II) was a more traditional FAR Part 15 fixed price incentive firm target contract by which the government and the contractor renegotiated the price and agreed to share cost growth up to a specified ceiling.[137] This was adopted because of the various challenges encountered and the need to mitigate Boeing's growing financial losses on the program's initial FFP contracts. In 2005, the SPO reported procurement cost, APUC, and schedule APB breaches for SVs 4 and 5 because of the need to allocate more funding based on the experiences with SVs 1 through 3.[138]

Ironically, WGS soon went through a very similar scenario a second time later in the program. As the severity of the technical and budgetary challenges facing TSAT continued to grow, doubts emerged about the viability of that program. Finally, in April 2009, DoD canceled TSAT. With AEHF experiencing delays and extreme cost growth, and TSAT canceled and

---

[135] DoD SAR, December 2002a, December 2003a, and DoD, *Selected Acquisition Reports: Wideband Global SATCOM (WGS)*, Washington, D.C., December 2004a.

[136] Marc Seilinger, "USAF: Launch of First WGS Delayed Due to Scheduling Problem," *Aerospace Daily and Defense Report*, March 22, 2005.

[137] Block II eventually included three WGS satellites. SV 6 was paid for by Australia in return for a share of the communication assets of the entire WGS constellation. See Julian Kerr, "Australia Commits to WGS Participation," *Jane's Defence Weekly*, October 3, 2006.

[138] DoD, *Selected Acquisition Reports: Wideband Global SATCOM (WGS)*, Washington, D.C., December 2005a.

WGS, which began as a temporary gapfiller and supplement to other, more advanced systems, shifted to a major long-term permanent component of DoD's future military satellite communications system. Its name was changed to WGS and, in recognition of this change, procurement of additional SVs—the B2FO—was approved. Yet, WGS program officials were confronted with the same dilemma that they experienced with Block II (SVs 4, 5, and 6). By the time a contract could be awarded for the WGS B2FO SVs, more than 24 months had passed since the exercise of the last Block II option, requiring negotiation of new contract terms and further price increases.

The B2FO initial contract was awarded in August 2010 and included options for WGS SVs 7 through 12. Full funding for B2FO SV 7 was not included until a contract modification in December 2011. Because of the large funding increase authorized for the B2FO SVs, the Air Force reported a critical APUC Nunn-McCurdy cost breach (27 percent) to Congress in March 2010, along with a significant schedule breach and increases in PAUC. In his letter to Congress, the Secretary of the Air Force explained that the breach was caused by the addition of SVs 7 and 8 to the FY 2011 President's Budget, and the greater-than-two-year production gap between SV 6 and SV 7. OSD completed the Nunn-McCurdy breach review and recertified the program to Congress in June 2010. According to the 2010 SAR, the main causes of the Nunn-McCurdy breach were "the two production breaks caused by the Government" and "the artificially low cost of the Block I satellites."[139] In other words, the SARs argue that the cost growth was principally caused by the two production gaps between Block I and Block II, and Block II and B2FO, as well as the overly optimistic cost estimates accepted by the government at MS B/C. Figure 2.8 shows the key events in the overall WGS program saga, including the two main production breaks that contributed to cost growth.

However, our assessment and other RAND research suggests that other factors contributed significantly to WGS Block II and B2FO cost growth. Critical among these were developmental and manufacturing problems, as well as changes in the commercial space market. A recent analysis of the program notes that moderate technical problems contributed to cost growth. These, in order of degreasing importance, included parts design and manufacturing difficulties, especially with the active-phased array antenna; launch vehicle integration issues; thermal vacuum test and transponder anomalies; B2FO software-development issues, and other technical problems. The two most important programmatic causes, according to the recent RAND analysis, were the added Block III (B2FO) and digital channelizer requirement; and the added Block II requirement.[140] This assessment makes clear, however, that although technical and other issues were important contributing factors, it was the restructuring of the program and added

[139] DoD, *Selected Acquisition Reports: WGS*, Washington, D.C., December 2010a, p. 4.

[140] Yool Kim, Elliot Axelband, Abby Doll, Mel Eisman, Myron Hura, Edward G. Keating, Martin C. Libicki, Bradley Martin, Michael McMahon, Jerry M. Sollinger, Erin York, Mark V. Arena, Irv Blickstein, and William Shelton, *Acquisition of Space Systems,* Volume 7: *Past Problems and Future Challenges*, Santa Monica, Calif.: RAND Corporation, MG-1171/7-OSD, 2015.

requirements brought about by delays in AEHF and cancellation of TSAT that were the main drivers of increased cost. Using the MS B quantity-adjusted PAUC and APUC, the program experienced little cost growth.

Figure 2.8. Time Line of Major WGS Events Showing Two Production Contract Gaps

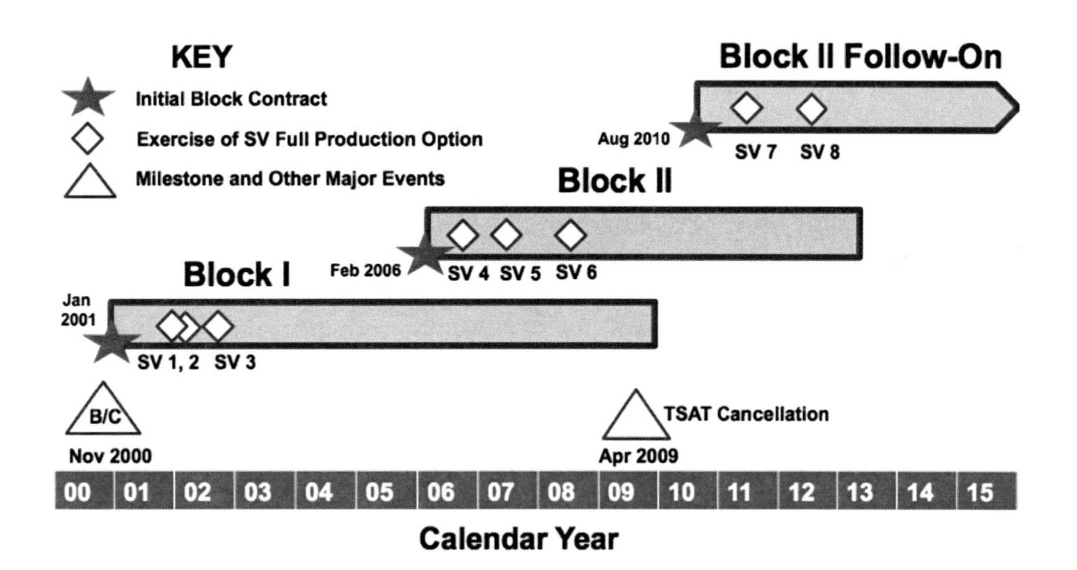

Given the history of the overall WGS program, which included nearly 50-percent APUC cost growth and a critical Nunn-McCurdy APUC breach, why was this program selected as one of four low cost-growth MDAPs? The following subsection provides a brief overview of how our cost methodology was applied to WGS.

Cost Methodology Applied to the WGS Program

A critical component of the RAND cost methodology used to determine accurate and comparable cost growth across many MDAPs is the normalization of cost-growth estimates to adjust for quantity or work-scope changes. Without doing this, cost-growth comparisons become distorted. Therefore, we normalize against the quantity planned in the original MS B APB. In the case of WGS, the original baseline envisioned the procurement of only three SVs. The major events described in the case study above that led to the Nunn-McCurdy breach and rise in costs were in large part outside the APB established at MS B and were caused by events outside the control of the SPO. The largest contributors to cost growth were the delays experienced by AEHF and the ultimate cancellation of TSAT, which led to the decision to expand production of WGS twice and necessitated contract renegotiation after production breaks.

The Air Force correctly points out that the cost estimates at MS B were moderately optimistic and contributed to later cost growth after Boeing experienced a variety of technical difficulties, particularly in manufacturing the early Block I SVs. However, this cost was not passed on to the government because the original first three SVs were procured under an FFP contract. Had the program never been expanded beyond the original planned three SVs or had the options for the three additional SVs been exercised within the 24-month contractual window, in theory Boeing would have had to absorb all the cost growth.[141] Prior to the addition of Block II satellites and the renegotiation of the production options, there was considerable cost growth in RDT&E, but virtually none in the normalized procurement price because of the FFP contract. Our methodology does not normalize RDT&E cost and, by the 2013 SAR, these costs had risen by more than 100 percent over the MS B baseline. However, the original MS B APB estimate for RDT&E was so small that this cost growth has had little effect on the overall program cost growth or APUC for the Block I SVs. In addition, at least some of this cost growth was due to work scope change, when DoD decided to triple the bandwidth available to accommodate RPV ISR vehicles.

In short, while RDT&E cost growth in percentage terms was not insubstantial, the original baseline estimate of RDT&E expenditure was so small that this cost growth, even if it had been passed on entirely to the government during Block 1, would have been quite modest and would not have changed our characterization of the Block 1 program as low cost growth. The real cost growth took place later, when the program was restructured, requirements increased, and the two extra blocks were added.

WGS Program Characteristics

The extensive case history of the WGS above alleviates the need for a lengthy discussion of the program characteristics. We offer summaries of the program characteristics as they apply to WGS Block 1.

Technology Maturity and Integration Complexity

WGS was planned at its inception to be a rapidly procured, low-cost, temporary solution to upgrade an aging satellite communications system pending the development and deployment of more-advanced next-generation satellites. For this reason, WGS was planned to use proven low-risk commercial technologies, components, processes, and procedures. The phased array antenna and the commercial satellite bus were the key components. However, the planning officials might be faulted for underestimating the manufacturing and integration challenges Boeing would experience with WGS, particularly with the phased array antenna and some system software.

---

[141] We recognize that many programs with FFP production contracts experiencing high cost growth for the contractor have led to the government agreeing to renegotiate the production contracts. This is typically done for industrial base reasons and an example of this is the T-6 Texan II primary trainer aircraft.

Requirements Realism and Stability

WGS requirements for Block I were reasonably realistic and stable. They may have implied some excessive optimism regarding the ease of manufacturing and integrating the commercial technologies (such as the phased array radar). Requirements stability and an FFP contract for Block I ensured low cost growth. However, over the entire life of the program, large quantity increases—combined with lengthy and costly unanticipated gaps between major production lots, as well as recovery from moderate technical problems and some increases in requirements and several other factors—were all important causes of cost growth in the APUC for Blocks II and B2FO satellites. The increased quantities were caused by the unanticipated requirement after Block I to replace the canceled TSAT constellation and delayed AEHF.

Cost-Estimate Realism

The MS B/C baseline cost estimate for the original Block I program was reasonably realistic. Risk may have been moderately underestimated in light of the various manufacturing problems encountered by Boeing, the dependence on future projections of a strong dual-use commercial market, and the problems experienced with such technologies as the active phased array antennas. Nonetheless, overall PAUC growth adjusted to the baseline numbers (Block I) grew only a very modest nine percent.

Acquisition Strategy and Program Structure

WGS focused heavily on a PBA-type procurement with strong emphasis on commercial technologies and approach. WGS Block I was procured under FAR Part 12 regulations for commercial-item procurement using prenegotiated FFP contracts. Although Boeing experienced RDT&E cost growth during Block I, this could not be passed on to the government because of the FFP production agreements. It would have been difficult for the SPO to anticipate such Boeing problems as the improper application of fasteners to SV 1 and the need to remove and replace them, which contributed to major launch delays. Like EELV, WGS also counted on a substantial commercial market, which failed to materialize. This caused supplier price increases and market exits leading to diminishing manufacturing sources.

Milestone B/C

WBS adopted a strategy of a combined MS B and MS C. This seemed reasonable at the time, given what was considered to be the commercial basis of the program and the need for minimal development with the use of FAR Part 12 commercial-item acquisition rules. This approach may have been unwise, however, given the many manufacturing challenges confronted by Boeing during the manufacture of the Block 1 SVs, particularly involving the phased-array antenna and the problems with improper application of fasteners. Most of these problems did not directly affect the price of the Block 1 SVs for the government because of prenegotiated fixed prices. Overall, these problems were relatively minor, with overall PAUC going up only 9 percent for Block I.

WGS Summary of Findings[142]

- Block I of the WGS program avoided most of the extreme cost-growth characteristics of the extreme cost-growth MDAPs by heavy reliance on commercial parts, processes, and technologies, but its FAR Part 12 commercial-item acquisition approach and FFP contract masked manufacturing and supplier problems, as well as market challenges.
- The WGS commercial-technology strategy appeared to work reasonably well for Block I in terms of actual PAUC growth, but the ease and cost of commercial-based technology manufacture and integration were underestimated, as were major supplier problems, and other issues.
- As with EELV, basing cost estimates on anticipated savings from commercial market demand was risky.
- Regarding the entirety of the WGS program and its three blocks, budgetary and unanticipated challenges caused by related programs increased quantity in an unforeseen way, causing major cost growth largely because of significant gaps between production lots that raised production for many reasons. Chief among these were the long delays (more than 24 months) between the blocks. This raised costs because of supplier issues and diminishing manufacturing sources caused in part by the failure of the commercial market to materialize, which caused the SV to evolve away from a commercial to more purely military systems. In addition, technical performance requirements were raised for the new SVs because of the cancellation of TSAT. Additional technical challenges were encountered on the new blocks. Finally, the Air Force permitted Boeing to raise its price because of losses experienced on the initial FFP Block I contract.

As Table 2.7 shows, WGS Block 1 had only one of the five key characteristics common to the six MDAPs with extreme cost growth and experienced only 9-percent growth in PAUC. When all blocks of WGS are included, cost growth at 48 percent is more significant. In line with our overall assessment, the WGS program—including all three procurement blocks—shares three characteristics with MDAPs with extreme cost growth, although not nearly to the same magnitude. The original WGS cost estimates at MS B were optimistic but, given the commercial approach to the original acquisition of three satellites and Boeing's acceptance of an FFP contract, the government experienced little PAUC growth for Block 1 satellites. However, the

[142] Another recent RAND study explicitly examining cost growth in space systems provides a more-detailed list of technical and programmatic issues that drove cost growth on the later blocks WGS (Block II and B2FO). These are consistent with what is noted here. We, however, are more interested in the Block I SVs that experienced low cost growth and that represent the original planned procurement. For the RAND analyses of the cost growth of later blocks, see Kim et al., 2015. pp. 19–23. For an earlier RAND assessment of the WGS Nunn-McCurdy cost breach, see Blickstein et al., 2011.

commercial FAR Part 12 commercial acquisition approach, which assumed WGS would remain close to a commercial communications satellite rather than evolving toward a military satellite, the unexpected need to procure many more SVs than originally intended, and the failure to exercise follow-on options from the original FFP contract within the contractually agreed 24 months, all contributed to more-significant cost growth for Blocks II and B2FO.[143]

Table 2.6. WGS Block I Had One of the Common Characteristics of the Six MDAPs with Extreme Cost Growth; All WGS Blocks Together Had Three

	WGS Block I Only (%)	WGS All Blocks (%)
Premature MS B		
Immature technology; integration complexity		
Unclear, unstable, or unrealistic requirements		
Unrealistic cost estimates		√
Acquisition policy and program structure		
Acquisition strategy and program structure not tailored for level of risk		√
MS B/C	√	√
PAUC growth (%)	9	**48**

NOTE: The light gray in the last column shows that, if the two follow-on buys are included (Block II and B2FO), more PAUC growth was experienced (a total of 48 percent). However, the two follow-on blocks are excluded in accordance with RAND methodology as explained in the text.

These issues suggest that the acquisition strategy and baseline cost estimates were generally appropriate for the originally planned procurement of only three SVs, but not appropriate for the much-expanded WGS program that emerged as TSAT became viewed as less and less viable.[144]

Chapter Three reviews what we learned about the four MDAPs with low cost growth and compares those programs with our six programs with extreme cost growth. The purpose is to validate the root causes of extreme cost growth and consider possible mitigation measures and approaches.

---

[143] In addition to many other factors, including parts design and manufacturing difficulties, such as with the phased-array antenna and some requirements growth, especially after the cancellation of TSAT. See Kim et al., 2015, for a detailed list of technical and programmatic issues. This analysis, however, confirms that none of these rose to the level of the technical and programmatic issues encountered on other recent space programs with extreme cost growth, such as AEHF and SBIRS.

[144] It could be argued that most of the factors that led to the more-significant cost growth on the full WGS program with all three procurement blocks were beyond the control of the Air Force and would have been difficult to anticipate.

3. Comparing Attributes of Low and Extreme Cost-Growth Programs

Introduction

As noted in Chapter One, the purpose of examining the detailed case histories of the four low cost-growth Air Force MDAPs was to compare their key characteristics with those of the six recent Air Force MDAPs with extreme cost growth, which we had previously examined. This was to verify that the key characteristics we identified in the six extreme cost-growth programs were indeed the likely root causes of extreme cost growth. Obviously, if the low cost-growth programs evinced characteristics similar to those found in the extreme cost-growth programs, we would need to rethink our assessment of the key characteristics of the extreme cost-growth programs with respect to root causes of extreme cost growth. Thus, the main questions we posed in our assessment of the four low cost-growth programs, as laid out in Chapter One, are as follows:

- Did these programs possess similar key characteristics and experience similar challenges as the extreme cost-growth programs? Why or why not?
- If so, how were they managed successfully?
- What else is different about these programs?

To reiterate, the key characteristics of the six extreme cost-growth programs we found in our earlier research can be summarized as shown in Table 3.1. These were divided into two main categories: (1) readiness for MS B and (2) acquisition policy and program structure. Essentially, the extreme cost-growth programs were not yet ready for MS B when they were approved for full-scale development. That is, there were too much known or unknown technology development and integration risk remaining; requirements had not yet been clearly defined with clear flow-down guidance; and, partly because of these factors, the MS B cost estimates were also too low, even though programs were often aware of independent cost estimates and the scale of the costs for prior generations of similar systems were considerably higher. Note that all six of the programs, with the exception of EELV, were characterized by all the elements composing the larger category of "Premature MS B." In addition, all six programs did not have acquisition strategies or program structures tailored for the level of risk. Four of the programs started with a combined MS B/C, which implies significant overlap of the development and production phases. This approach may work on some very low-technology risk, commercially based acquisitions, but for most programs, it is a high-risk approach that can lead to LRIP and further production with unstable designs because of continuing RDT&E. The latter can lead to the need for costly retrofits and upgrades later in the production phase.

Table 3.1. Two Categories of Common Characteristics of Six MDAPs with Extreme Cost Growth

	AEHF	C-130 AMP	EELV	Global Hawk	NPOESS	SBIRS High
Premature MS B						
Immature technology; integration complexity	√	√		√	√	√
Unclear, unstable, or unrealistic requirements	√	√		√	√	√
Unrealistic cost estimates	√	√	√	√	√	√
Acquisition policy and program structure						
Acquisition strategy and program structure not tailored for level of risk	√	√	√	√	√	√
MS B/C	√		√	√	√	
PAUC growth (%)	95	193	273	152	154	279

NOTES: The bottom row shows PAUC growth for each program. There is little or no correlation between the number of characteristics evident in a specific program and the severity of that program's cost growth in percentage terms. Each MDAP is unique in context and circumstances, and this is why it is so important to convey the details of each case history and ultimately to compare these six "worst-of-the-worst" cases to best-performing MDAPs.

Findings and Observations

Our assessment of the four programs with low cost growth raises two basic questions:

- Do these findings tend to confirm that the key characteristics of the six programs with extreme cost growth are indeed the key cost drivers of extreme cost growth?
- What is the applicability of the acquisition approaches used on the four programs with low cost growth to future programs?

The Root Causes of Extreme Cost Growth

Our findings regarding the four programs with low cost growth tend to support the contention that the key characteristics identified on the programs with extreme cost growth in our prior research are indeed important causes of extreme cost growth and cost growth in general. The low cost-growth programs exhibit few or none of the key characteristics of the six extreme cost-growth programs, as shown in Table 3.2.[145]

The C-5 RERP is an anomaly. It shows two of the six characteristics of the extreme cost-growth programs, including underestimation of technical complexity and difficulty, and unrealistic initial cost estimates. This is perhaps to be expected because, at 18 percent, it experienced by far the most PAUC growth of the four programs with low cost growth. WGS Block I shows one of the characteristics of the extreme cost-growth programs, and it had the second-highest cost growth of the four low cost-growth programs. And, if all three blocks are

[145] See, also Chapter Three, Table 3.1, and Chapter One, Table 1.2.

Table 3.2. Four Low Cost-Growth MDAPs Have Less of the Common Characteristics of the Six MDAPs with Extreme Cost Growth

	C-5 RERP	JDAM	SDB I	WGS Block I Only	WGS All Blocks
Premature MS B					
Immature technology; integration complexity	√				
Unclear, unstable, or unrealistic requirements					
Unrealistic cost estimates	√				√
Acquisition policy and program structure					
Acquisition strategy and program structure not tailored for level of risk					√
MS B/C				√	√
PAUC growth (%)	18	–12	–16	9	**48**

NOTES: The last row shows PAUC growth for each program per our calculations. The light gray in the last column notes that if the two follow-on buys are included (Block II and B2FO), more PAUC growth was experienced (a total of 48 percent). However, RAND cost methodology excludes these two follow-on blocks as explained in the text.

examined, it experienced more cost growth, although not nearly as much as programs with extreme cost growth.

The remaining two programs, JDAM and SDB I, reveal none of these characteristics, and they both had negative cost growth. (In other words, their final PAUC was actually substantially lower than estimated at MS B.) Furthermore, the severity and magnitude of the issues confronted by C-5 RERP regarding technology complexity and cost estimates, and by WGS Block I with respect to premature MS C, were of far less magnitude and seriousness than those experienced on the six programs with extreme cost growth. Indeed, the combining of the development and production phases, the overly optimistic and small RDT&E funding, the award of FFP contracts for the first three SVs, and the award of the contract under FAR Part 12 commercial-item acquisition regulations, all show that the government considered WGS to be essentially equivalent to procurement of a COTS item. In a similar manner, the most expensive component of the C-5 RERP program was the purchase of a commercial derivative engine from GE. Much of the cost growth on RERP was due to outside factors, such as the cost growth and delays experienced on the C-5 AMP baseline system development and production.

The final outcome of the entire WGS program, including all blocks, also seems to confirm this interpretation. As was noted earlier in Chapter Two in the WGS case study, the RAND methodology for normalizing for quantity changes, when comparing MS B cost estimates with later estimates, compares the later estimate adjusted for the same quantity as planned at MS B. This addresses the problem of underestimating cost growth when programs later cut quantities to reduce overall cost when there is significant cost growth.

In the case of WGS, the original baseline assumed funding for only three satellites. While the original contract had options for three more satellites, these options were not initially funded and, when finally exercised, they were beyond the contractually required period. Thus, given the original baseline estimate and assumptions, the first three SVs were procured with only minimal PAUC growth. However, if the two follow-on buys are included (Block II and B2FO), more PAUC growth was experienced (a total of 48 percent), as shown in light gray in Table 3.2.

Furthermore, if we include all the blocks in our assessment of the entire program, we can see that the program also suffered from two other characteristics common to the extreme cost-growth programs: unrealistic cost estimates at MS B and an acquisition strategy and program structure not tailored to the level of risk. Again, however, these challenges were not nearly of the same magnitude as similar challenges confronted by the programs with extreme cost growth. Therefore, it should not be surprising that the WGS program taken as a whole (including all blocks) experienced more significant cost growth, but not dramatically more than is typical of the average Air Force MDAP and far lower that the programs with extreme cost growth.

In short, our assessment of four MDAPs with low cost growth and comparison with our earlier analysis of six MDAPs with extreme cost growth indicate that there is indeed a strong correlation between the program attributes that we have identified in our prior research and extreme cost growth. Conversely, these attributes are found to be almost entirely lacking, or of much smaller magnitude, in the programs with low cost growth. Therefore, this comparison provides us with much greater confidence that the key characteristics of programs we identified in our earlier research are indeed among the key drivers of extreme cost growth on recent Air Force MDAPs.

The second question we list at the beginning of this section raises the issue of how applicable these findings are to most future Air Force MDAPs, which is related to the question of how the programs with low cost growth differ in significant ways, if at all, from the programs with extreme cost growth and in ways other than those involving the five characteristics of programs with extreme cost growth. We turn to these issues in the next subsection.

Applicability of Findings to Future Air Force MDAPs

Are there other significant attributes common to all the four low cost-growth programs that clearly differentiate them from all the programs with extreme cost growth and thus can be associated with the difference in cost-growth outcomes? Indeed, every MDAP is unique to a lesser or greater degree in both internal and external circumstances.

Referring back to Tables 1.4 and 1.5 in Chapter One, which provide a cost overview of all the programs, it is apparent that most of the programs with extreme cost growth are much larger in dollar value, both in total program costs and RDT&E costs. However, this could be partly because of the substantial cost growth experienced by the extreme cost-growth programs compared with the low cost-growth programs. How do these two groups of programs compare in dollar value at MS B when the original baseline cost estimates were formulated?

The baseline program cost estimates for all ten programs are shown in Figure 3.1. Here, we see that, with the exception of the C-5 RERP, the original program estimates at MS B suggest the low cost-growth programs tend to be generally smaller in terms of cost value compared with the extreme cost-growth programs, but the differences compared with Tables 1.4 and 1.5 after cost growth are less pronounced. Not only is the C-5 RERP well within the range of the extreme cost-growth programs at the MS B estimate, but JDAM is also approximately equal to NPOESS, which experienced extreme cost growth, while JDAM experienced negative cost growth. Program cost estimates at MS B thus do not appear to distinguish sufficiently among programs with different cost growth.

Figure 3.1. Total Program Cost Estimates at MS B, by Cost-Growth Category

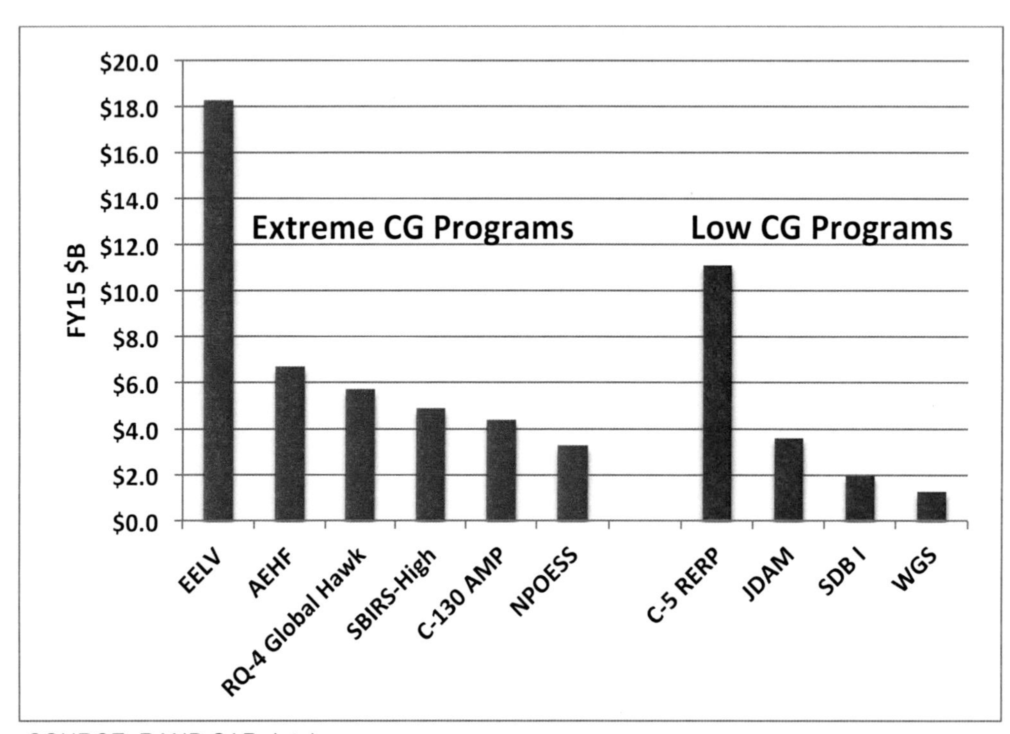

SOURCE: RAND SAR database.
NOTE: CG = cost growth.

The picture changes significantly if we examine estimated RDT&E costs for all ten programs at MS B. Here, we see a more-pronounced difference between the programs with extreme cost growth compared with those with low cost growth, as shown in Figure 3.2.

Figure 3.2 compares the MS B RDT&E cost estimates of all the MDAPs with extreme cost growth and low cost growth, as well as the December 2013 SAR cost estimates in constant 2015 dollars, showing actual cost growth to that point. Here, all of the programs with low cost growth have smaller RDT&E estimates at MS B, with the exception of the C-5 RERP. Figure 3.3 includes only the MS B RDT&E estimates for all the programs, showing, even more clearly, how low the original MS B RDT&E estimates were for the low cost-growth programs compared with the extreme cost programs. For the overall program estimates at MS B (Figure 3.1), the C-5 RERP estimate is higher than all but the highest MDAP in the group of extreme cost-growth programs, yet it is the MS B estimate for RDT&E (Figures 3.2 and 3.3) that is smaller than the estimates for one-half of the programs with extreme cost growth and roughly equal to that of EELV, which is the third-smallest estimate for RDT&E at MS B for the extreme cost-growth programs. Furthermore, the other three programs with low cost growth have estimates considerably below the lowest MS B RDT&E estimate for extreme cost-growth programs (C-130 AMP). In addition, the best performing of the low cost-growth programs—JDAM and SDB I—had RDT&E estimates at MS B far below any of the extreme cost-growth programs. The worst-performing of the programs with low cost growth—C-5 RERP—had by far the highest RDT&E estimate at MS B and indeed experienced total PAUC growth of 18 percent, far more than the other low cost-growth programs and not that far off from the 95-percent PAUC growth experienced by AEHF, as shown in Chapter One, Tables 1.3 and 1.4.

Figure 3.2. RDT&E Cost Estimate at MS B and December 2013 by Cost-Growth Category

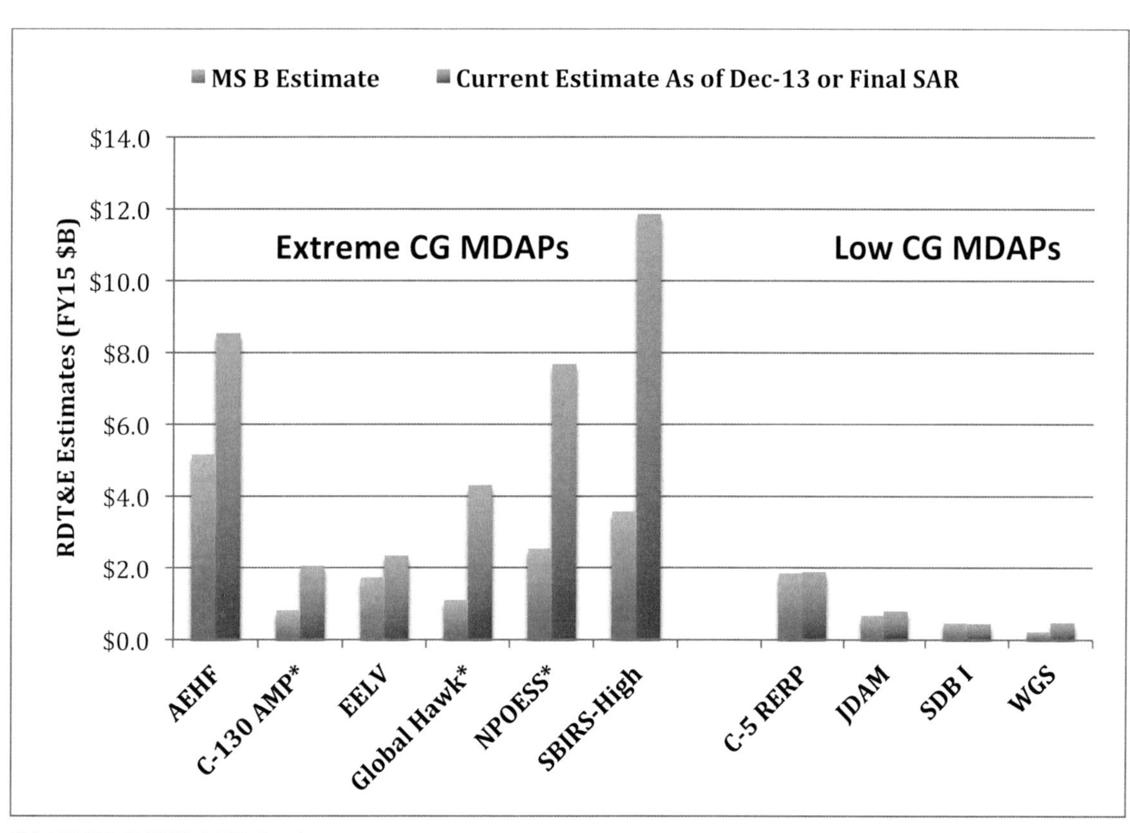

SOURCE: RAND SAR database.
NOTES: CG = cost growth.
* Programs were terminated or truncated in the FY 2013 President's Budget.
Note that three of the four programs with low cost growth have MS B estimates considerably below the lowest MS B RDT&E estimate for an extreme cost-growth program.

In addition, as shown on Table 1.1 in Chapter One and Figure 3.4, most of the extreme cost-growth programs experienced a high percentage of RDT&E budgetary cost growth measured from MS B, with four programs in triple digits. In stark contrast, the low cost-growth programs experienced low or negative cost growth in budgetary RDT&E from MS B, with the exception of WGS, as shown in Table 1.3 in Chapter One and Figure 3.4. In the case of WGS, with all blocks included, the relatively high cost growth is partly an artifact of RAND's cost methodology, which makes no normalization adjustments for quantity changes or SOW changes in RDT&E costs. As noted earlier, we include only the WGS Block I satellites in our estimate of low cost-growth programs, because this was the original intent of the program. Thus, RDT&E cost growth for WGS Block I and WGS Blocks I, II, and B2FO are shown separately in Figure 3.4. During the initial phases of the program, prior to the decision to buy additional SVs and the lengthy and costly production gaps between Block I, Block II, and B2FO, WGS experienced relatively low cost growth in RDT&E, as also shown in Figure 3.4.[146]

Figure 3.3. RDT&E Estimates at MS B (FY15 $B)

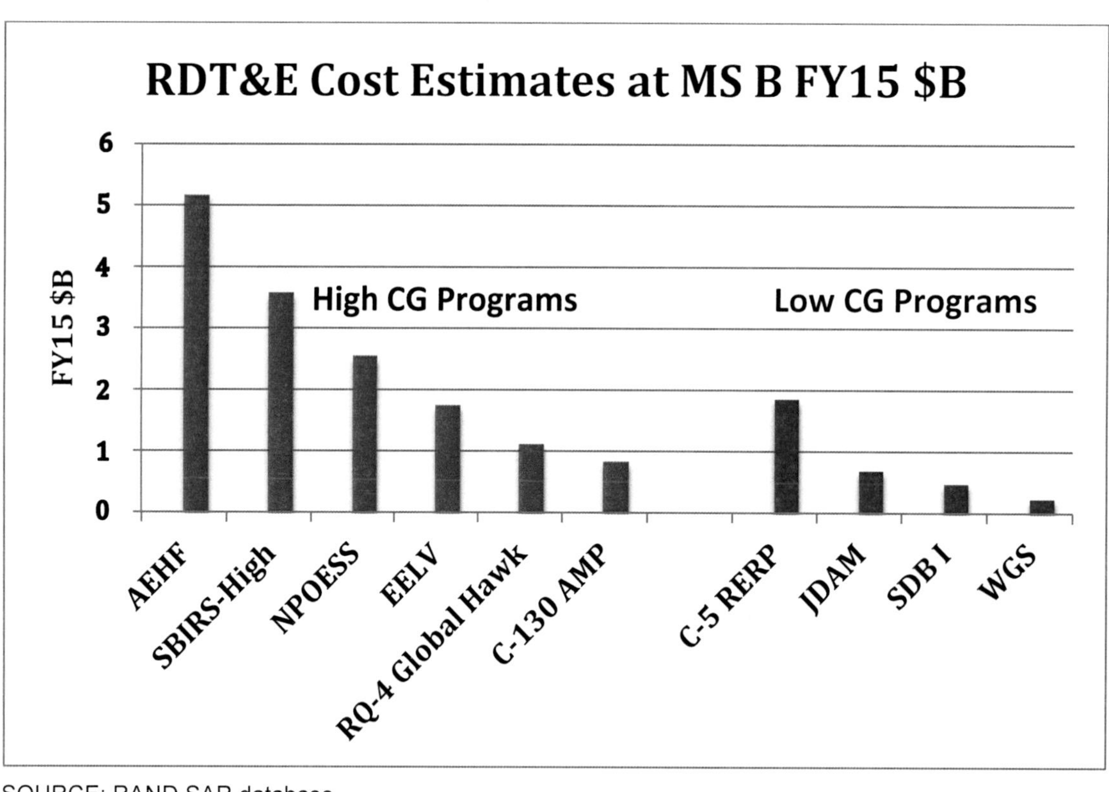

SOURCE: RAND SAR database.
NOTE: CG = cost growth.

[146] Overall, however, including Blocks I, II, and B2FO, WGS experienced 113-percent cost growth in RDT&E, as shown in Figure 3.3. In addition to other factors mentioned, this was also due to the assumption at MS B that the original Block I satellite would be virtually COTS satellites, which resulted in a very small and unrealistically low initial RDT&E budget estimate.

Figure 3.4. Percentage Budgetary RDT&E Cost Growth By Cost-Growth Category

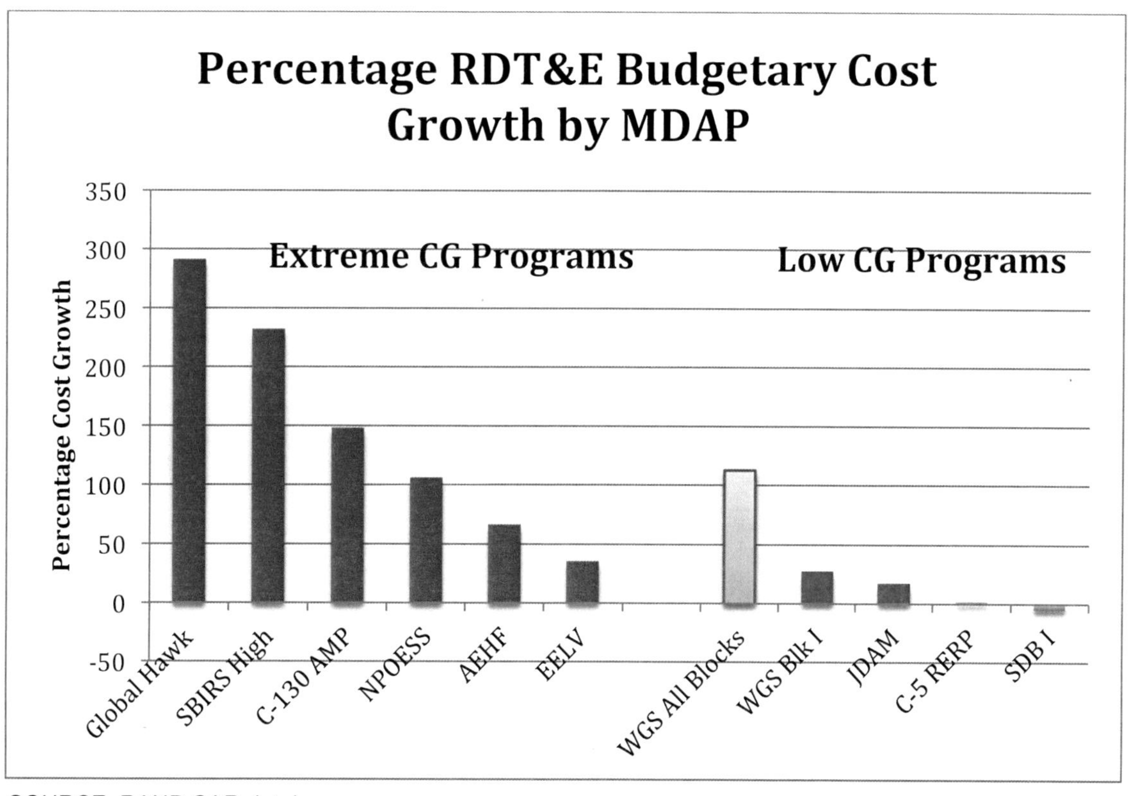

SOURCE: RAND SAR database.
NOTES: The pink bar shows RDT&E cost growth for the entire WGS program, including Blocks I, II, and B2FO. We treat only WGS Block I as a low CG program, as explained in Chapter Two. Cost growth is measured from MS B through FY 2013. CG = cost growth.

What can we deduce from this comparison? The six extreme cost-growth programs had higher RDT&E estimates at MS B—generally much higher—than the low cost-growth programs. Furthermore, the extreme cost-growth MDAP RDT&E cost estimates at MS B were obviously underestimated, since most of the extreme cost-growth programs experienced high cost growth in RDT&E, while, for the most part, the low cost-growth programs experienced modest or even negative cost growth in RDT&E. The size of the initial RDT&E cost estimates and their subsequent performance regarding cost growth, might be viewed as a surrogate for technological and integration complexity and difficulty.

Thus, the six extreme cost-growth programs tended to be much larger, more complex, and more challenging development and production programs than the four programs with low cost growth. This observation is generally confirmed by closer examination of the detailed case studies. This is particularly true when the best-performing programs, JDAM and SDB I, are looked at in more detail. The JDAM tail-kit assembly contains a relatively low number of key subsystems and subcomponents, most of which are derivative of mature, commercially developed technologies and components. The most important are likely the guidance control unit, which contains the laser gyro INS, the GPS receiver, and the mission computer.[147] The overall complexity and difficulty of the development and production effort would seem to be less challenging than those of many MDAPs, such as SBIRS and NPOESS. This is illustrated by the fact that JDAM became essentially a commodity item, produced in quantities almost of the magnitude of commercial items (more than 40 per day), amounting by 2014 to well in excess of 220,000 kits.[148] A parallel argument can be made for SDB I.

In addition, all four of the low cost-growth MDAPs were intended to be highly dependent on COTS and GOTS technology. Indeed, procurement of the original WGS SVs was conducted under the provisions of FAR Part 12 commercial item acquisition rules, implying it was equivalent to purchasing an existing commercial item. WGS had a small initial RDT&E budget as a result. In a similar manner, the single most-expensive aspect of the C-5 RERP program was the GE engine, which was a direct commercial derivative of the commercial GE CF6-80C2 engine. Both JDAM and SDB I—besides being relatively less complex—heavily emphasized use of commercial and commercially derived or GOTS components and subsystems. In addition, they were designated as formal DAPPs and thus received high priority from OSD and Air Force leadership and broad regulatory relief from costly reporting and oversight requirements.

In contrast, such programs as SBIRS and NPOESS, contain many hundreds, if not thousands, of complex military-unique components, subsystems, and technologies. Design, development, and integration of advanced sensors were particularly challenging on the extreme cost-growth programs. Even shortly after MS B, before the extreme cost growth was recognized, the SBIR's APUC was estimated at more than $248 million (FY 1995 BY$) per SV.[149] In the case of NPOESS, the original APUC was estimated at more than $242 million (2002 BY$) per SV.[150] This compares with an APUC of $19,000 (FY 1995 BY$) per tail kit for JDAM.[151] Clearly, development—and particularly integration—of such extremely complex systems as SBIRS High and NPOESS are far more demanding and inherently more risky than a relatively less-complex system, such as JDAM or SDB I.

We have found that the four MDAPs experiencing low cost growth have relatively low dollar value, heavy use of derivative mature commercial technologies, and generally low complexity. Does this commonality of characteristics limit the applicability of any inferences we might draw? Does it mean that the lessons learned from the low cost-growth programs are only truly

---

[147] For one open-source technical description, see Carlo Kopp, "JDAM Matures Part 1 and 2," *Airpower Australia*, updated January 27, 2014.

[148] DoD, 2009–2014 (2014).

[149] DoD, *Selected Acquisition Reports: Space-Based Infrared System High (SBIRS High)*, Washington, D.C., 1997.

[150] DoD, *Selected Acquisition Reports, National Polar-Orbiting Operational Environmental Satellite System (NPOESS)*, Washington, D.C., 2002.

[151] However, the WGS communications satellite initial APUC estimate, at $265 million 2001 BY$, was comparable to the original SBIRS and AEHF APUC estimates.

applicable to similar types of lower-complexity commercially derivative programs? Not necessarily. Our reasons are as follows.

Extreme Cost Growth and Possible Mitigation Approaches

We argue that the key characteristics of the programs with extreme cost growth, as summarized in Table 3.1, represent the root causes of extreme cost growth, and apply to all MDAPs no matter how complex and challenging. The key characteristics of the extreme cost-growth programs are nearly absent from the low cost-growth programs (with those present of a much smaller magnitude), as shown in Table 3.2. This confirms our hypothesis from our earlier research that these characteristics are important root causes of extreme cost growth. What the lower complexity of the low cost-growth programs tells us is that it is easier to apply strategies and approaches to mitigate these characteristics in less-complex programs than in more-complex programs. However, we maintain that a low complexity program will not necessarily automatically achieve low cost growth without application of mitigating measures, particularly in the area of technology and integration risk, or in the application of high-level interest.[152]

Our findings that low cost-growth programs tend to be less challenging and complex structurally and technologically supports our central recommendation for mitigating extreme cost growth in our earlier research.[153] Our two key recommendations from the earlier report were to

- ensure that programs have realistic cost estimates at MS B
- embrace evolutionary or incremental strategies with comprehensive and proven implementation strategies.

Extensive discussions of our more-detailed recommendations are viewable in our earlier report, and are not repeated here. Two points bear repetition and even greater emphasis, however, based on the findings of this report.

While the larger, more-complex programs with extreme cost growth were all characterized by unrealistic and overly optimistic cost estimates at MS B, the programs with low cost growth tended to have far more realistic cost estimates at MS B. Nearly all the extreme cost-growth programs experienced substantial cost growth in RDT&E, which spilled over into even greater cost growth during the procurement phase. While it could be argued that it is easier to develop realistic cost estimates for less-complex programs or programs based on commercial systems, this argument is not compelling. In nearly every case, the extreme cost-growth programs were aware of much more realistic cost estimates from independent sources, as well as from the experience with prior programs in similar types of systems. More realistic estimates at MS B

---

[152] Both JDAM and SDB I benefited enormously from their special status as high-visibility DAPPs and the special attention and prioritization they received from senior Air Force officials.

[153] For a complete discussion of the findings and recommendations of our earlier research, see Lorell, Leonard, and Doll, 2015.

were clearly possible, and while they might have made approval and budgeting for the programs more difficult, they would have resulted in far less cost growth and program and budgeting disruption down the road. Interestingly, the two lower cost-growth programs with the most cost growth (C-5 RERP and WGS) also suffered from moderately unrealistic cost estimates at MS B, but of a much smaller magnitude than in the case of the extreme cost-growth programs. It is crucial that the Air Force ensure that realistic cost estimates based on a clear understanding of the developmental and production risks entailed in the program are developed at MS B to avoid excessive and practically inevitable cost growth later.

We believe that the second key recommendation in our earlier report stressing the importance of properly applied evolutionary or incremental acquisition strategies is even more relevant in light of the research findings reported here. In essence, most of the extreme cost-growth programs' problems stemmed from a gross underestimation of the complexities and uncertainties confronting them in designing, developing, integrating, and producing very challenging technological systems. One way to mitigate this problem may be to reduce the complexity of challenging systems by breaking them down into more manageable phases or increments rather than pursuing the very high-risk approach of seeking a single step to full capability. Incremental or evolutionary acquisition is difficult to plan and implement, but the payoffs can be high.[154] Nonetheless, there may be times when the warfighter needs a new high-risk advanced capability as soon as possible, and there is not time for incremental or evolutionary acquisition. There could be another time when the advanced high-risk attributes of the new system are not readily amenable to divide into more manageable, lower-risk increments. In such cases, it may, under some circumstances, be justified to pursue a higher-risk approach of single step to full capability.[155] Yet, even in such cases, a much more realistic assessment and estimate of the potential costs, the technology risks, and the known and unknown unknowns, are possible and highly desirable to avoid extreme cost growth and the budgetary and programmatic disruptions caused by programs that experience extreme cost growth. We urge the Air Force to continue to focus on improving the accuracy and reliability of the cost estimating process, particularly during early program phases.

---

[154] Determining the viability of and best implementation strategies for incremental acquisition are areas deserving of much more extensive analysis. The views expressed in this report are based on the analysis of the case studies, as well as early RAND research published in Mark A. Lorell, Julia F. Lowell, and Obaid Younossi, *Evolutionary Acquisition: Implementation Challenges for Defense Space Programs*, Santa Monica, Calif.: RAND Corporation, MG-431-AF, 2006.

[155] Some observers allege that such programs do not bring advanced capability to the warfighter any faster and thus are *never* justified.

Bibliography

Arena, Mark V., Robert S. Leonard, Sheila E. Murray, and Obaid Younossi, *Historical Cost Growth of Completed Weapon System Programs*, Santa Monica, Calif.: RAND Corporation, TR-343-AF, 2006. As of August 11, 2015: http://www.rand.org/pubs/technical_reports/TR343.html

Assistant Secretary of Defense for Research and Engineering, *Technology Readiness Assessment (TRA) Guidance*, Washington, D.C., U.S. Department of Defense, April 2011.

Blickstein, Irv, Michael Boito, Jeffrey A. Drezner, James Dryden, Kenneth Horn, James G. Kallimani, Martin C. Libicki, Megan McKernan, Roger C. Molander, Charles Nemfakos, Chad J.R. Ohlandt, Caroline Reilly, Rena Rudavsky, Jerry M. Sollinger, Katharine Watkins Webb, and Carolyn Wong, *Root Cause Analyses of Nunn-McCurdy Breaches*, Volume 1: Zumwalt-*Class Destroyer, Joint Strike Fighter, Longbow Apache, and Wideband Global Satellite*, Santa Monica, Calif.: RAND Corporation, MG-1171/1-OSD, 2011. As of December 15, 2016: http://www.rand.org/pubs/monographs/MG1171z1.html

Boeing, "Backgrounder: Small Diameter Bomb Increment I (SDB) I," January 2012. As of December 5, 2016: http://www.boeing.com/assets/pdf/defense-space/missiles/sdb/docs/SDB_overview.pdf

Bolten, Joseph G., Robert S. Leonard, Mark V. Arena, Obaid Younossi, and Jerry M. Sollinger, *Sources of Weapon System Cost Growth: Analysis of 35 Major Defense Acquisition Programs*, Santa Monica, Calif.: RAND Corporation, MG-670-AF, 2008. As of August 11, 2015: http://www.rand.org/pubs/monographs/MG670.html

Butler, Amy, "Wideband Gapfiller Launch Raided to Pay for FY02 SBIRS High Bailout," *Inside Missile Defense*, August 21, 2002.

———, "Wideband Gapfiller Satellite to Launch Nearly a Year Late," *Inside Defense*, August 1, 2003.

C-5 Reliability Enhancement and Re-engining Program (RERP), *Selected Acquisition Reports*, December 2001–2014.

Chien, Daniel, *Ready or Not? Using Readiness Levels to Reduce Risk on the Path to Production*, Falls Church, Va.: General Dynamics, August 2011.

Committee on Earth Studies, Space Studies Board, Commission on Physical Sciences, Mathematics, and Applications, National Research Council, *The Role of Small Satellites in NASA and NOAA Earth Observation Programs*, Washington, D.C.: National Academy Press, 2000.

Department of Defense Instruction 5000.2, *Operation of the Defense Acquisition System*, Washington, D.C.: U.S. Department of Defense, May 12, 2003.

Department of Defense Instruction, 5000.02, *Operation of the Defense Acquisition System*, Washington, D.C., U.S. Department of Defense, January 7, 2015.

DoD—*See* U.S. Department of Defense.

DoDI—*See* Department of Defense Instruction.

Fox, J. Ronald, *Defense Acquisition Reform: 1960–2009: An Elusive Goal*, Washington, D.C.: Center of Military History, U.S. Army, 2011.

Friar, Allen, *Cost Growth and the Limits of Competition*, Huntsville, Ala.: Defense Acquisition University, September 2012.

GAO—*See* U.S. Government Accountability Office.

Gertler, Jeremiah, *Air Force C-17 Aircraft Procurement: Background and Issues for Congress*, Washington, D.C.: Congressional Research Service, 7-5700, RS22763, December 22, 2009. As of May 20, 2015:
https://fas.org/sgp/crs/weapons/RS22763.pdf

Imbens, Guido W., and Donald B. Rubin, *Causal Inference in Statistics, Social, and Biomedical Sciences*, New York: Cambridge University Press, 2015.

Jennings, Gareth, "Lockheed Martin Progresses C-5M Galaxy Deliveries to USAF," *IHS Jane's 360*, November 4, 2014.

Joint Requirements Oversight Council, "C-5 Reliability Enhancement and Re-Engining Program Nunn McCurdy Certification: Supporting Explanation," January 2008.

Kerr, Julian, "Australia Commits to WGS Participation," *Jane's Defence Weekly*, October 3, 2006.

Kim, Yool, Elliot Axelband, Abby Doll, Mel Eisman, Myron Hura, Edward G. Keating, Martin C. Libicki, Bradley Martin, Michael McMahon, Jerry M. Sollinger, Erin York, Mark V. Arena, Irv Blickstein, and William Shelton, *Acquisition of Space Systems,* Volume 7: *Past Problems and Future Challenges*, Santa Monica, Calif.: RAND Corporation, MG-1171/7-OSD, 2015. As of December 15, 2016:
http://www.rand.org/pubs/monographs/MG1171z7.html

Knight, William, and Christopher Bolkom, *Strategic Airlift Modernization: Analysis of C-5 Modernization and C-17 Acquisition Issues*, Washington, D.C.: Congressional Research Service, RL34264, October 22, 2008. As of August 12, 2015: http://digital.library.unt.edu/ark:/67531/metacrs10719/m1/1/high_res_d/RS22763_2008Oct22.pdf

Kopp, Carlo, "JDAM Matures Part 1 and 2," *Airpower Australia*, updated January 27, 2014.

Leonard, Robert S., and Akilah Wallace, *Air Force Major Defense Acquisition Program Cost Growth Is Driven by Three Space Programs and the F-35A: Fiscal Year 2013 President's Budget Selected Acquisition Reports*, Santa Monica, Calif.: RAND, RR-477-AF, 2014. As of August 11, 2015: http://www.rand.org/pubs/research_reports/RR477.html

Lorell Mark A., and John C. Graser, *An Overview of Acquisition Reform Cost Savings Estimates*, Santa Monica, Calif.: RAND Corporation, MR-1329-AF, 2001. As of August 12, 2015: http://www.rand.org/pubs/monograph_reports/MR1329.html

Lorell, Mark A., Robert S. Leonard, and Abby Doll, *Extreme Cost Growth: Themes from Six U.S. Air Force Major Defense Acquisition Programs*, Santa Monica, Calif.: RAND Corporation, RR-630-AF, 2015. As of December 5, 2016: http://www.rand.org/pubs/research_reports/RR630.html

Lorell, Mark A., Julia F. Lowell, Michael Kennedy, and Hugh P. Levaux, *Cheaper, Faster, Better? Commercial Approaches to Weapons Acquisitions*, Santa Monica, Calif.: RAND Corporation, MR-1147-AF, 2000. As of August 12, 2015: http://www.rand.org/pubs/monograph_reports/MR1147.html

Lorell, Mark A., Julia F. Lowell, and Obaid Younossi, *Evolutionary Acquisition: Implementation Challenges for Defense Space Programs*, Santa Monica, Calif.: RAND Corporation, MG-431-AF, 2006. As of January 24, 2017: http://www.rand.org/pubs/monographs/MG431.html

Office of the Under Secretary of Defense for Acquisition, Technology, and Logistics, *Report of the Defense Science Board/Air Force Scientific Advisory Board Joint Task Force on Acquisition of National Security Space Programs,* Washington, D.C., May 2003.

"Pentagon Recertifies Air Force C-5 Re-Engining Program," *Inside the Air Force*, February 22, 2008.

Porter, Gene, Brian Gladstone, C. Vance Gordon, Nicholas Karvonides, R. Royce Kneece Jr., Jay Mandelbaum, and William D. O'Neil, *The Major Causes of Cost Growth in Defense Acquisition*, Vol. I, *Executive Summary*, Alexandria, Va.: Institute for Defense Analyses, IDA Paper P-4531, December 2009a.

———, *The Major Causes of Cost Growth in Defense Acquisition*, Vol. II, *Main Body*, Alexandria, Va.: Institute for Defense Analyses, IDA Paper P-4531, December 2009b.

Public Law 87-653, Truth in Negotiations Act, September 10, 1962.

Sauser, Brian J., and Jose Ramirez-Marquez, *System (of Systems) Acquisition Maturity Models and Management Tools*, Hoboken, N.J.: Stevens Institute of Technology, 2009.

"Saving the Galaxy: The C-5 AMP/RERP Program," *Defense Industry Daily*, June 18, 2015. As of August 11, 2015: http://www.defenseindustrydaily.com/saving-the-galaxy-the-c-5-amprerp-program-03938/

Seilinger, Marc, "USAF: Launch of First WGS Delayed Due to Scheduling Problem, Job Cuts," *Aerospace Daily and Defense Report*, March 22, 2005.

"Senators Press Air Force Brass on Airlift Recapitalization Needs," *Inside the Air Force*, March 23, 2007.

"Statement of Sue C. Payton, Assistant Secretary of the Air Force for Acquisition, U.S. Air Force, Accompanied by Diane M. Wright, Deputy Program Executive Officer for Aircraft, Aeronautical Systems Center, Wright-Patterson Air Force Base," in *Cost Effective Airlift in the 21st Century*, hearing before the Federal Financial Management, Government Information, Federal Services, and International Security Subcommittee of the Committee on Homeland Security and Governmental Affairs, U.S. Senate, 110th Congress, 1st Sess., Senate Hearing 110–410, September 27, 2007, pp. 8–10 and 49–57. As of August 12, 2015: http://www.gpo.gov/fdsys/pkg/CHRG-110shrg38845/pdf/CHRG-110shrg38845.pdf

"USAF Directed to Find $1.8 Billion for C-5M, Consider Logistics Support," *Inside the Air Force*, May 5, 2008.

U.S. Code, Title 10, Section 2433, Unit Cost Reports (UCRs), January 7, 2011.

U.S. Code of Federal Regulations, Title 32, Section 2.4, Designation of Participating Programs: National Defense, Washington, D.C.: U.S. Government Publishing Office, July 1, 2001.

U.S. Department of Defense, *Selected Acquisition Reports: Space-Based Infrared System High (SBIRS High)*, Washington, D.C., 1997.

———, *Selected Acquisition Reports: Wideband Global Satellite System (WGS)*, Washington, D.C., December 2000.

———, *Selected Acquisition Reports: Wideband Global SATCOM (WGS)*, Washington, D.C., December 2001a.

———, *Selected Acquisition Reports: Wideband Global SATCOM (WGS)*, Washington, D.C., December 2002a.

———, *Selected Acquisition Reports: Wideband Global SATCOM (WGS)*, Washington, D.C., December 2003a.

———, *Selected Acquisition Reports: Wideband Global SATCOM (WGS)*, Washington, D.C., December 2004a.

———, *Selected Acquisition Reports: Wideband Global SATCOM (WGS)*, Washington, D.C., December 2005a.

———, *Selected Acquisition Reports: Wideband Global SATCOM (WGS)*, Washington, D.C., December 2006a.

———, *Selected Acquisition Reports: Wideband Global SATCOM (WGS)*, Washington, D.C., December 2007a.

———, *Selected Acquisition Reports: Wideband Global SATCOM (WGS)*, Washington, D.C., December 2008a.

———, *Selected Acquisition Reports: Wideband Global SATCOM (WGS)*, Washington, D.C., December 2009a.

———, *Selected Acquisition Reports: WGS*, Washington, D.C., December 2010a.

———, *Selected Acquisition Reports: WGS*, Washington, D.C., December 2011a.

———, *Selected Acquisition Reports: Wideband Global SATCOM (WGS)*, Washington, D.C., December 2012a.

———, *Selected Acquisition Reports: Wideband Global SATCOM (WGS)*, Washington, D.C., December 2013a.

———, *Selected Acquisition Reports: Wideband Global SATCOM (WGS)*, Washington, D.C., December 2014a.

———, *Selected Acquisition Reports (SAR)*: *C-5 Reliability Enhancement and Re-Engining Program (C-5 RERP)*, Washington, D.C., 2001b.

———, *Selected Acquisition Reports (SAR)*: *C-5 Reliability Enhancement and Re-Engining Program (C-5 RERP)*, Washington, D.C., 2002b.

———, *Selected Acquisition Reports (SAR)*: *C-5 Reliability Enhancement and Re-Eengining Program (C-5 RERP)*, Washington, D.C., 2003b.

———, *Selected Acquisition Reports (SAR)*: *C-5 Reliability Enhancement and Re-Engining Program (C-5 RERP)*, Washington, D.C., 2004b.

———, *Selected Acquisition Reports (SAR)*: *C-5 Reliability Enhancement and Re-Engining Program (C-5 RERP)*, Washington, D.C., 2005b.

———, *Selected Acquisition Reports (SAR): C-5 Reliability Enhancement and Re-Engining Program (C-5 RERP)*, Washington, D.C., 2006b.

———, *Selected Acquisition Reports (SAR): C-5 Reliability Enhancement and Re-Engining Program (C-5 RERP)*, Washington, D.C., 2007b.

———, *Selected Acquisition Reports (SAR): C-5 Reliability Enhancement and Re-Engining Program (C-5 RERP)*, Washington, D.C., 2008b.

———, *Selected Acquisition Reports (SAR): C-5 Reliability Enhancement and Re-Engining Program (C-5 RERP)*, Washington, D.C., 2009b.

———, *Selected Acquisition Reports (SAR): C-5 Reliability Enhancement and Re-Engining Program (C-5 RERP)*, Washington, D.C., 2010b.

———, *Selected Acquisition Reports (SAR): C-5 Reliability Enhancement and Re-Engining Program (C-5 RERP)*, Washington, D.C., 2011b.

———, *Selected Acquisition Reports (SAR): C-5 Reliability Enhancement and Re-Engining Program (C-5 RERP)*, Washington, D.C., 2012b.

———, *Selected Acquisition Reports (SAR): C-5 Reliability Enhancement and Re-Engining Program (C-5 RERP)*, Washington, D.C., 2013b.

———, *Selected Acquisition Reports (SAR): C-5 Reliability Enhancement and Re-Engining Program (C-5 RERP)*, Washington, D.C., 2014b.

———, *Selected Acquisition Reports, National Polar-Orbiting Operational Environmental Satellite System (NPOESS)*, Washington, D.C., 2002c.

———, *Selected Acquisition Reports: Small Diameter Bomb I (SDB-I)*, Washington, D.C., December 2003c.

———, *Selected Acquisition Reports: Small Diameter Bomb I (SDB-I)*, Washington, D.C., December 2004c.

———, *Selected Acquisition Reports: Small Diameter Bomb I (SDB-I)*, Washington, D.C., December 2005c.

———, *Selected Acquisition Reports: Small Diameter Bomb I (SDB-I)*, Washington, D.C., December 2006c.

———, *Selected Acquisition Reports: Small Diameter Bomb I (SDB-I)*, Washington, D.C., December 2007c.

———, *Selected Acquisition Reports, Joint Direct Attack Munition (JDAM)*, Washington, D.C., 2004d.

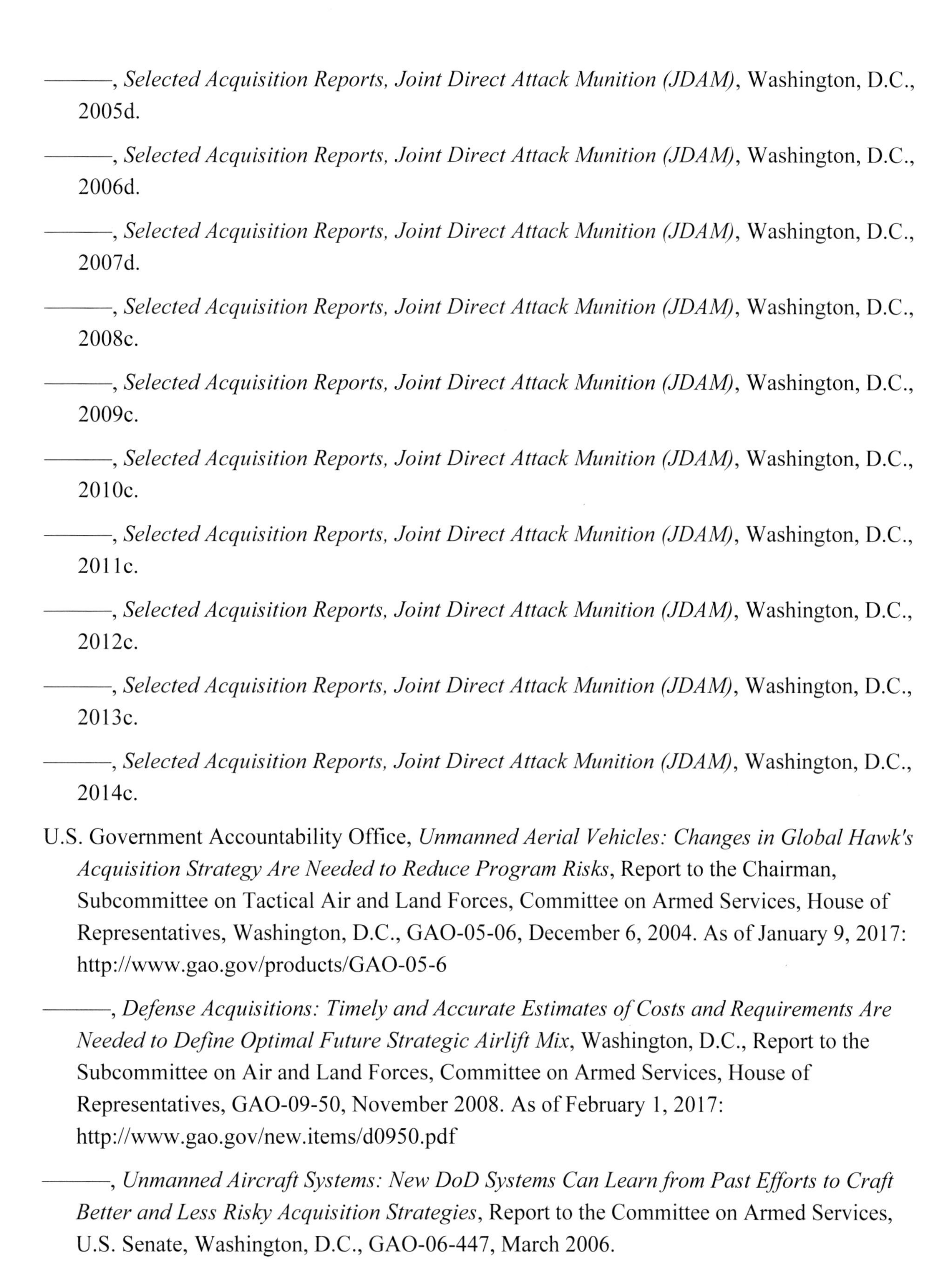
———, *Selected Acquisition Reports, Joint Direct Attack Munition (JDAM)*, Washington, D.C., 2005d.

———, *Selected Acquisition Reports, Joint Direct Attack Munition (JDAM)*, Washington, D.C., 2006d.

———, *Selected Acquisition Reports, Joint Direct Attack Munition (JDAM)*, Washington, D.C., 2007d.

———, *Selected Acquisition Reports, Joint Direct Attack Munition (JDAM)*, Washington, D.C., 2008c.

———, *Selected Acquisition Reports, Joint Direct Attack Munition (JDAM)*, Washington, D.C., 2009c.

———, *Selected Acquisition Reports, Joint Direct Attack Munition (JDAM)*, Washington, D.C., 2010c.

———, *Selected Acquisition Reports, Joint Direct Attack Munition (JDAM)*, Washington, D.C., 2011c.

———, *Selected Acquisition Reports, Joint Direct Attack Munition (JDAM)*, Washington, D.C., 2012c.

———, *Selected Acquisition Reports, Joint Direct Attack Munition (JDAM)*, Washington, D.C., 2013c.

———, *Selected Acquisition Reports, Joint Direct Attack Munition (JDAM)*, Washington, D.C., 2014c.

U.S. Government Accountability Office, *Unmanned Aerial Vehicles: Changes in Global Hawk's Acquisition Strategy Are Needed to Reduce Program Risks*, Report to the Chairman, Subcommittee on Tactical Air and Land Forces, Committee on Armed Services, House of Representatives, Washington, D.C., GAO-05-06, December 6, 2004. As of January 9, 2017: http://www.gao.gov/products/GAO-05-6

———, *Defense Acquisitions: Timely and Accurate Estimates of Costs and Requirements Are Needed to Define Optimal Future Strategic Airlift Mix*, Washington, D.C., Report to the Subcommittee on Air and Land Forces, Committee on Armed Services, House of Representatives, GAO-09-50, November 2008. As of February 1, 2017: http://www.gao.gov/new.items/d0950.pdf

———, *Unmanned Aircraft Systems: New DoD Systems Can Learn from Past Efforts to Craft Better and Less Risky Acquisition Strategies*, Report to the Committee on Armed Services, U.S. Senate, Washington, D.C., GAO-06-447, March 2006.

———, *Defense Acquisitions: Assessments of Selected Weapons Programs*, Report to Congressional Committees, Washington, D.C., GAO-10-388SP, March 2010a. As of August 12, 2005:
http://www.gao.gov/assets/310/302379.pdf

———, *Defense Acquisitions: Strong Leadership Is Key to Planning and Executing Stable Weapon Programs*, Report to the Committee on Armed Services, U.S. Senate, Washington, D.C., GAO-10-522, May 2010b.

———, *Defense Acquisition Process: Military Service Chiefs' Concerns Reflect Need to Better Define Requirements Before Programs Start*, Report to the Honorable James Inhofe, U.S. Senate, Washington, D.C., June 2015a. As of December 6, 2016:
http://www.gao.gov/assets/680/670761.pdf

———, *Evolved Expendable Launch Vehicle: The Air Force Needs to Adopt an Incremental Approach to Future Acquisition Planning to Enable Incorporation of Lessons Learned*, Report to Congressional Committees, Washington, D.C., GAO-15-623, August 2015b.

U.S. Senate, 110th Cong., 1st Sess., Cost Effective Airlift in the 21st Century, Hearing Before the Federal Financial Management, Government Information, Federal Services, and International Security Subcommittee of the Committee on Homeland Security and Government Affairs, Washington, D.C.: U.S. Government Printing Office, Senate Hearing 110-410, September 27, 2007.

"Wideband Gapfiller System," GlobalSecurity.org, undated. As of August 6, 2015:
http://www.globalsecurity.org/space/systems/wgs.htm

Witek, Roger J., *Hard Skills, Soft Skills, Savviness and Discipline—Recommendations for Successful Acquisition: Case Studies of Selected Boeing Weapons Programs*, Montgomery, Ala.: Maxwell Air Force Base, Air University, AU/AFF/013/2008-05, May 2008.